FACTORS AFFECTING
DISPERSAL DISTANCES OF SMALL ORGANISMS

Factors Affecting Dispersal Distances of Small Organisms

D. O. Wolfenbarger

An Exposition-University Book

Exposition Press   **Hicksville, New York**

To Dr. F. M. Wadley

FIRST EDITION

© 1975 by D. O. Wolfenbarger

LIBRARY OF CONGRESS CATALOG CARD NUMBER: 73-92856

ISBN 0-682-47905-5

Printed in the United States of America

Contents

Preface

Interest, the desire for knowledge, and the need of understanding dispersal distances of small organisms motivated the efforts expended in the making of this book. Factors of various kinds affect distances to which small organisms disperse or are dispersed. What are these factors and how much do they increase or decrease distances of movement?

Such distances refer to field situations or natural conditions that have value or give understanding for the benefit of man. In order to measure or to know what these factors are and the magnitude of their effects, prodigious efforts were required. Only by assiduous efforts and the contributions of many people could the achievement of even a minor goal be expected. Examination of the literature of the world, in as far as posssible, has provided bases for the contents. References to all of such literature, however, cannot be expected, especially to those published in faraway places and in diverse languages.

It is understandable that only part of the publications presenting dispersal distance observations would have material to construct a discipline. A minimum of three distance points with population sample parameters, for example, was chosen as a standard for inclusion in this book. Separation of the species factor from the other factors presented herein, was necessary because of the volume of material. The presentation of that material and more recent references may be given in a later publication.

I am indebted to Dr. F. M. Wadley for his encouragement and influence in dispersal studies and to members of the Hume Library of the University of Florida for help to me without which this volume could not have been prepared.

Introduction: Concepts of Dispersion

Organisms employ varied, many and often ingenious methods in their continual and essential activities to occupy new areas. Bacteria, fungus spores, seed, pollen and insects all disperse in order to perpetuate their kind. They are constantly dispersing to areas in which the species is absent or less populous, although organisms tend to concentrate in locations favorable to their growth and development. Concentration of organisms decreases as the distance from their point of origin increases (Andrewartha, 1961). Thus, interesting gradient patterns are established. An increased understanding of these patterns would help explain certain ecological phenomena and also aid in solving some economic problems facing man. There is a need for further understanding concerning dispersion distances and some factors that aid or retard dispersion distance. Previous reviews by Wolfenbarger (1946, 1959) provide a broad viewpoint of dispersion and a foundation on which to base this presentation. Many definitive observational details of microorganisms dispersion have been provided by Gregory (1961), many insects by Johnson (1969).

Most references cited pertain to macroscopic organisms, since large organisms may be observed and sampled more readily than smaller ones. The dispersion of microscopic organisms, however, is of much significance. For example, it has been found that bacteria are carried by medical laboratory personnel. In hospitals, personnel, equipment, and air movement (cross ventilation) are agents moving bacteria. Infectious organisms that disperse or are dispersed by agents decrease with distance and increase with time from the contaminant source.

Many references contain only qualitative information about the dispersion of organisms. This may suffice where there is no need or desire to know the dispersion pattern. Practical application requires more definitive and objective information—quantitative data, especially the distances from the source to which organisms move or are moved.

Very little data of this nature is in the literature, and it is difficult to obtain (Godwin, 1934). More dispersion references have appeared

during the last decade, however, than in previous years. Certain outstanding researchers have tended to stimulate the study and understanding of dispersion. Noteworthy among these are Gressitt (1961), Moulton
(1942), Pijl (1969), Schneider (1962), Schwerdtfeger (1942), and Willis
(1922).

CONCEPTS OF DISPERSION

Two objects cannot occupy the same space at the same time; two
individuals of the same species must have two geographical spaces.
Progeny, as a rule, move to one side of the parent and must disperse
to occupy space, develop to maturity, and reproduce the species.

Dispersion affects both the species involved and other species. Advantages of dispersion for the species involved include: (1) aid in entering habitats new to the species; (2) more complete utilization of habitat
resources; (3) reduction of inbreeding; (4) crossing, with resultant hy-
brid vigor; (5) promotion of uniform hereditary factors throughout a
population; and (6) promotion of the spread of new genes. Another con-
sequence of dispersion is competition for or reduction of space and food
for some species and increased food and shelter for others. Dispersal is
also an important factor in maintaining the population equilibrium and/
or alternatively in eliminating the destructive inter- and intra-specific
competitions (Itô, 1960). For man some beneficial effects of dispersion
include food production and shelter through the scattering of seeds and
pollen of many fruit, fiber, and vegetable crop plants.

Deleterious effects of dispersion include (1) the spread of organisms,
causing many plant and animal diseases; (2) contamination of food;
(3) competition for food and space; and (4) the cost of control efforts.
It is well known that destructive organisms can and do spread rapidly
once they gain a foothold in favorable locations where inadequate or
no control measures are applied. Examples of this are the Mediterranean
fruit fly in Florida and hoof and mouth disease elsewhere. Both have
been eradicated only at tremendous expense. Strict quarantines are main-
tained to prevent, in so far as possible, entry of such pests into this
country.

Most studies of the dispersion of organisms are made in response to
or in connection with some economic need of man. Dispersion that bene-
fits man may, however, either benefit or injure other organisms.

Two types of movement among organisms may be recognized—dispersion and migration. Dispersion may be the more primitive, but some
species employ both types of movement, depending on the stage and
phase of activity. The term *migration* is used more correctly with refer-

ence to fish and waterfowl movements than with movements of bacteria, fungi, or insects (Thomson, 1929). The general spreading of an organism from one place to another is called *dispersion* unless migration has been proved to be the habit of the organism.

Terminology and concepts. Dispersion is the dissemination, emanation, exodus, movement, or scattering of organisms from their source. The word *dispersion* is (1) generally accepted by usage, (2) applicable by definition, and (3) expressive of the process involved. It is applied here to movements of a species, variety, or group through one generation, phase, stage, cycle or activity. When a species from an established area enters a new locality, it disperses. (Dispersion in reference to the variability or spread of statistical units is a different concept from that used here.)

It is not always possible to distinguish between the primary and secondary spread of organisms. With regard to the species about which they are knowledgeable, most workers can be reasonably certain of whether primary or secondary spread is involved and whether it has significant magnitude. Organisms originating from secondary or later cycles initiate a new dispersion process, which must be recognized as separate from the primary movement.

Migration. Many large and colorful insects—butterflies, for example —moving in the same direction in large numbers, as if on parade, are very striking in appearance. Mass movements, or swarms, of locusts also are striking, because of the catastrophic results from their destruction of food crops. Such movements may be called *migrations*. Migrations are characterized by continued movement in a more or less general direction, with both the movement and general direction under the control of the animal concerned, and there may or may not be a "return flight" (Clark, 1931; Williams 1958). Odum's (1959) concept of migration was periodic departure and return; Tutt (1902) referred to it as "irregular dispersal movements and by no means regular movements to and from a given locality."

Common (1954) describes an outstanding example of insect migration, that of the bogong moth, *Agrotis infusa* (Boisd). McFadden (1941) describes the dispersion of wheat stem rust spores, *Puccinia graminis* Pers., as an example of migration of a passive disperser organism; the spores move northward from Texas and other southern areas in the spring and then in the fall move southward, where the species may remain active during winter.

Distribution. Dispersion and distribution involve separate biological activities. Permanent residence of a species in a given place is usually considered a prerequisite of distribution, but periodical or occasional residence may be accepted as a suitable condition. A species living in an area only occasionally, or seldom, however, approaches migratory status. Except to distinguish it from dispersion, no attempt is made here to discuss the wide and general subject of geographical distribution. Although "dispersion" was used in the title of a book by Wynne-Edwards (1962), he was actually discussing principally distribution and spatial relationships. Occasional occupancy, density, scatter, "property tenure," and holding sites may be given biological consideration under certain conditions.

Some dispersal characteristics. Most small organisms move about aimlessly, wandering willy-nilly or mechanically and apparently without definite purpose or goal, over paths that may cross and recross. Some organisms that control their movements are often lured to particular areas or locations.

Other organisms may move about with a purposeful movement. The honeybee, *Apis mellifera* (L.), is an outstanding example of this type. Besides flying purposely towards a source of nectar, the honeybee communicates with its colony members as to distance, direction, quantity, and kind of floral source (von Frisch, 1950). *Purposive dispersion* is recognized as that in which organisms move during different stages, phases, or activities in reaction to situations meeting such needs as obtaining food, protection, mates, swarming, or egg deposition.

Each species has one or more stages, phases, or activities in which dispersion may or does occur. For example, the orange, *Citrus sinensis* Osbeck, has three disperse phases: pollen, fruit (seed), and nursery trees. These characteristically have species differentiations for distances. In the salt marsh mosquito, *Aedes taeniorhynchus* (Wied.), all life stages (eggs, larvae, pupae, and adults) and several phases of adult activities (swarming, feeding, mating, and egg deposition) are disperse phases or stages. Each stage, phase, or activity may have a different dispersal gradient and range from others of the same species.

Many theories have been developed to explain the motivation of insect dispersion. Frankel (1932) groups the motivational causes this way:

1. Periodicity or rhythmic cycles
2. Hunger
3. Propagation

One or more of these motivations may be operative in the dispersion of all active disperser organisms. Periodical movements and hunger may be absent with a given species. Propagation is a potent motive common to all species and is probably the most important one. Another factor, crowding, may be added, although in many instances motivations of hunger and propagation may be the same as crowding.

Active and passive dispersion. Two types of dispersion are recognized, active and passive, depending on the source of energy that transports the organism. When transportation is supplied by energy from within the organism, the dispersion is active; when, supplied by energy outside the body, the dispersion is termed passive. Bacteria, fungi, pollen, and seed are usually considered passive disperser organisms, although some bacteria and fungi are motile and disperse actively. The movement of pollen, seeds, and spores when sprouting might also be said to be active dispersion, but the process does not rightly belong in the dispersion category.

Means or agencies of dispersion. Increased understanding of the distances to which organisms move (or are moved) may be obtained by understanding the various means or agencies by which dispersion occurs. Thre is usually a principal mode of transportation for each species. Most organisms disperse, or may be dispersed, by more than one mode or agency. This is a form of insurance for the species, since the limitation of passage to a single agency might prove harmful or disastrous.

Everything that possesses motion may be an agent that disperses small organisms. Heald *et al.* (1915) and Gardner (1918) dealt with the agencies and methods of the dispersion of fungi and discussed some of the relationships and problems involved. A later and more complete listing presented by Wolfenbarger (1946) is now modified and recast as follows:

SOURCES OF ENERGY, TRANSPORTERS OR MODES OF MOVEMENT

A. Movement of passive organisms:
    1. Wind or air currents, especially turbulences
    2. Rain or dew, atmospheric moisture
    3. Water currents, rivers, oceans, lakes, ponds irrigation, water conduits, sewage
    4. Earthquake, land slide or other geologic disturbance
    5. Other organisms (than man for which a special listing is made)
B. Movement of active organisms:
    1. Flying
    2. Crawling, creeping
    3. Walking, running
    4. Jumping
    5. Swimming
C. Propulsion:
    1. Ejecting
    2. Dehiscing

D. Movement by man as the agent—products, articles of commerce:
1. Fruits, fibers
2. Seeds, nuts
3. Vegetating organisms—roots, tubers, corns, bulbs rhizomes, cuttings
4. Nursery stock, scions, seedlings
5. Soil, litter, compost, scrap iron, ballast
6. Crude products, hay, straw, grain (feed), packing material
7. Manufactured or processed goods
8. Forest products, lumber, logs
9. Stone, masonry, or masonry products
10. Mine quarry, ore, or quarry products
11. Agricultural operations—planting, cultivating, harvesting, processing, milling, milled products.

E. Movement by man's carriers:
1. Airplane
2. Automobile
3. Barges
4. Boats, rafts
5. Foot traffic
6. Projectiles, rockets, missiles
7. Railways
8. Ships
9. Truck, trailer
10. Vehicular—horse-drawn or other

Transportation of organisms by man is one of the principal modes of dispersion into previously or recently uninfested areas. A brief discussion of the dispersal of mosquitoes through human transportation was published by Hughes and Porter (1956). Although the authors made reference to dispersal as "accidental," such movements remain dispersion whether "unintentional," "purposeful," or any other term is used, and whether by the oldest or latest modes of transportation. Only one group of arthropods, mosquitoes, was discussed by Hughes (1961), and principally the immature stages of the organism. Other groups of organisms that might have been discussed are those affecting man through agricultural crops, animals, clothing, food, forest wood products, or other areas. Perhaps man's studies of the more severe pests should be more abundant.

Models or patterns of dispersal. Patterns or models of the dispersion of organisms are helpful for understanding. Patterns of *Tilletia* and *Bovista* spore dispersal were presented by Stepanov (1935), *Cercospora* spores by Öort (1936), *Drosophila* flies by Dobzhansky and Wright

(1943) and Timofeeff-Ressovsky (1940, 1940a, 1940b) and others have also illustrated the dispersal of organisms. The author presents a figure of a laboratory dispersion range in some tests (fig. 1). In this procedure insects were released at a central point to disperse omnidirectionally. Annuli were drawn on the dispersion range to indicate distances. Although examinations of all areas in each annulus were made for organisms, samples or portions could be taken of parts of the annuli. Lines were drawn to divide the dispersion range into compass quadrants to determine any directional influence.

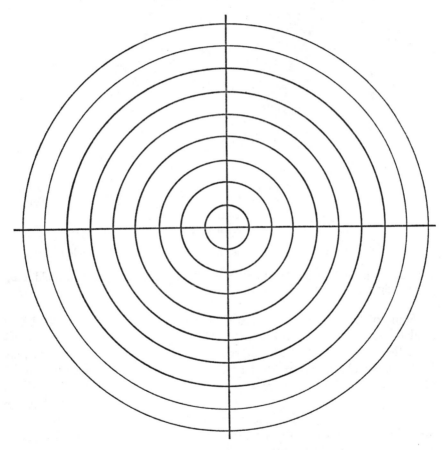

Fig. 1. A simple experimental design, floor-plan or dispersion range to measure dispersion. Annuli centering at the origination site extend concentrically to distances from the center, depending on the organism to disperse whether in terms of mm or of hundreds of km from the source. Radii extending cardinally from the center indicate directional effects or provide a replicate feature.

A three-dimensional model is presented to illustrate the dispersion of many field crop infestations (fig. 2.). Severest attacks are seen at the corner where plants are attacked by flies from two directions. Attacks are shown to decrease towards midfield.

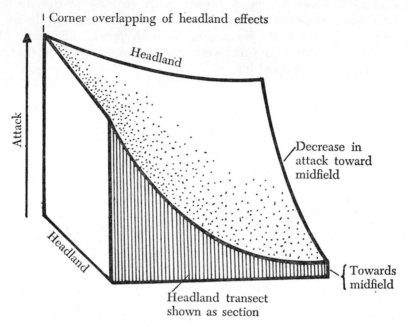

Pictorial diagram of attack on the corner of a field

Fig. 2.   Model of an invasion exemplified for many pests of field crop plants (fig. 2, Wright and Ashby, 1946).

Some of the regression curves shown by Wolfenbarger (1946, 1959) tend to be exponential in character (fig. 3); more seem to have, however, the quarter part of an ellipsoidal pattern. Exponential curves may be the more elemental or ideal type of curve. The proximal segment of dispersed organisms is more separable than the intermedial or distal zones. Organisms in the proximal zone, younger initially than those in the other zones, may disperse more rapidly and efficiently than after they disperse intermedially or distally, due to age. Apparent lengthening or flattening of the distal segment may result from more fortuitous movements or less hazardous journeys than the others. Movements from unknown or outside sources of organisms may enter the dispersion range and mix with those in the distal segments more frequently than in the proximal or intermediate zones. Sources of organisms in the distal zones may be considered background areas.

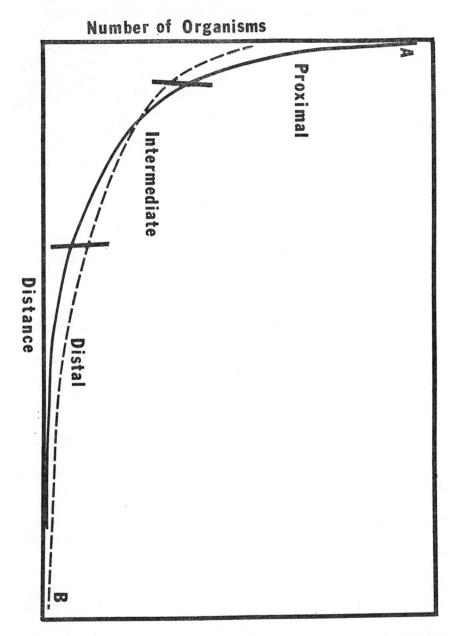

Fig. 3. Trends in regression curves: Exponential pattern A and one-quarter part of an ellipsoidal pattern B. Both curves may be considered as having proximal, intermedial and distal segments, depending on distances from the source of organisms.

Short- or long-distance dispersion. Dispersion acts were classified as "continuous" or "discontinuous" by Butler (1917), "regular" or "sporadic" by Fletcher (1925). Terms such as "short" and "long" distances and similar terminology have been employed to express dispersion and incidence. Obviously, these are purely relative terms, the meaning of which depends on one of several factors. More meaningful terminology would appear to include "usual," "frequent," "infrequent," "probable," or "improbable." More definitive data, however, are desired in order to prevent misinterpretation. Obviously, distances between individuals, or spacing, are involved between organisms, whatever factors may be affecting dispersion.

SAMPLING CONSIDERATIONS

It is practical to sample dispersing populations but impractical to count all the individuals, unless only a few are involved. Fewer organisms are usually found as distances increase from the origin; they decrease rapidly at first, then less rapidly. Graphically, most regression curves depicting dispersion tend to approach, but not to reach, zero. Some samples of a population, however, may include zero observations of frequency. Uncertainties exist, indicating that zero is reached at some unknown distance from the origin. Long "tails" are usually indicated by graphic studies if extensive data are available.

Since it is almost impossible to discover the maximum distance of dispersal, an alternative is needed. Such an alternative may be the recognition of *distances to which part or most organisms disperse.* Control or other measures could not be adapted with most organisms to utilize the total distance dispersed even if it were known.

Dyeing, painting, dusting, tagging and hybridizing of individuals have been used to differentiate the dispersion phases of organisms from wild or uncontrolled populations so that the dispersants might be recognized. Tagging organisms with radioactive materials is a comparatively recent innovation that appears to be very suitable for measuring dispersion.

Choice of a method or procedure to use in studying dispersion depends on a number of factors, such as the species, stage, phase, and activity of the organism and conditions at hand, in connection with the objective sought. In any extensive study some knowledge of the biology of the organism involved and of sampling procedures is needed. Some conception of the distance range involved is also desirable. Some species, stages, phases, and activities disperse in distances of inches. In other instances the distance ranges are measured in hundreds of miles. Sampling procedures must be extensive enough to obtain adequate numbers and yet small enough to remain within practical budgetary limits.

Dispersion of small organisms usually refers to movement in the field where there is little or no control of the movements by man. Such dispersion presumably occurs under "natural" conditions and is the result of net responses to various factors. Although research conducted under field conditions may be less exact than that under controlled or laboratory conditions, it follows free movement.

Statistical measurement and determination. Quantitative determinations obtained by some methodical sampling procedures listed above have included counts at different distances from the source of organisms. Comparisons are thus made of the relative abundances from which gradients are obtained to illustrate the dispersion rates. Statistical methods in common use are employed to reduce to more comprehensive form and to evaluate the masses of data obtained from sampling procedures. Among the many authors who have discussed statistical procedures are: Bateman (1947, 1950), Braun *et al.* (1953), Burla *et al.* (1950), Bitancourt and Fawcett (1944), Fracker (1936), Frampton *et al.* (1942), Gregory (1945), Johnson (1969), Hansing and Frampton (1940), Ingold (1956), Kettle (1951), Wadley and Wolfenbarger (1944), Wadley (1957), Wilson and Baker (1946a), Wright and Ashby (1946), and Zentmyer *et al.* (1943). These citations discuss statistical methods and formulae used in dispersal measurement and determination. This researcher found (Wolfenbarger, 1946, 1959) that no one regression formula would function satisfactorily with all data. Different formulae, moreover, could often be used for the same data without loss of any important relationship. Formulae and concepts discussed by Wadley (in Wadley and Wolfenbarger, 1944) have been applied herein as they were also used by Wolfenbarger (1946, 1959).

Zonal divisions in regression curves. An idealized relationship between the amount of pollen rain and the distance from the source was discussed by Janssen (1966). Although no data were given, a graph was used to show the relative amounts of pollen falling at various distances. Three classifications, or *zones,* were described. Since they are characteristic of many or even most organismal dispersal curves (Wolfenbarger, 1946, 1959), they may be considered in dispersion studies. These zones were termed "local," "extralocal," and "regional" (Fig. 3). Organisms from nearby, or local, sources have rapid rates of *regression,* as shown by rapid falling off, of pollen rain with initial distance increases from the local source. *Transitions,* longer differences, occur over intermediate distances, recognized as the *extralocal* zone. This zone occurs as the sharp turn, or "elbow," in the regression curve. Pollen abundance was found to be distinctly lower in the extralocal than in the local zone but was higher than at longer distances. At greater distances, however, the pollen dispersal curve tends to parallel the abscissa. In this, the *regional* zone, amount of pollen was least of any zone and changed but little over longer distances. The pollen rain at these greater distances reportedly originated from regional vegetation.

Regional, or *"background,"* numbers of organism (units or incidence numbers, ratios or percentages) are frequently observed in dispersion studies if distances from a known source are sufficiently great. Such regional data are doubtless of much significance, although they are not well understood at present.

Studies of regression by Wadley and Wolfenbarger (1944) showed that semilogarithmic formulae were the most suitable way of expressing the relationships. The formulae were developed where the

$$\text{Expected} = a + b \,(\log x), \text{ or}$$
$$\text{Expected} = a + b \,(\log x) + c \, l/x.$$

The symbols *a, b,* and *c* are factors obtained from the data, as in regular regression statistical calculations, and are used for determining the slope of the curve. The *x* refers to distance.

Other formulae, as indicated by Gregory (1961), are often used and serve well to indicate the trend of movement. These include the following:

$$\text{Log of expected values} = a + bx, \text{ and}$$
$$\text{Log of expected values} = \log a + bx.$$

These or other formulae sometimes fail to illustrate apparently faithful trends of dispersal shown by individual sets of data. This seems especially true where only three to five distance classes with their component dispersal attributes are given. Dispersion studies illustrating this problem include celery mosaic by Wellman (1935), potato calico by Porter (1935), Douglas-fir stocking by Hofmann (1911), pea weevil dispersion by Wakeland (1934), and blow-fly dispersion by Schoof and Siverly (1954). Sometimes an exponential curve indicates a characteristic curve, as if it were fundamental in dispersion, especially if many data were taken over very wide distance ranges. This is exemplified by the studies of walnut pollen dispersal by Crane *et al.* (1937), *Simulium* sp., fly dispersal by Dalmat (1950), and subarctic mosquitoes recovered by Jenkins and Hassett (1951). Again, in most instances, an exponential regression does not indicate the regression of organisms on distance.

Gross or net distance. Consideration of two factors is needed for more complete understanding of dispersal distance. One factor is the total distance, or the sum of all movements. This might be termed *gross distance* and would include all meanderings. It is seldom measured, owing prehaps to the laborious details attendant to its accomplishment. Certain objectives may be obtained, however, by measuring total movements. The other factor is the net, or radial, *distance* from the origin, which is the one usually employed. Net distance is the more important and more practical consideration.

Some species may be found with little or almost no difference between gross and net dispersal distance. Passive disperser organisms may initiate and terminate dispersion with straight rather than labyrinthine or winding movements. Active disperser organisms usually conduct more tortuous paths during dispersal journey than do passive disperser species; therefore, wider differences are expected between the gross and net distances dispersed by active disperser organisms.

Radial or areal relationships. Each investigator may question whether to use radial or areal units to determine regression relationship of data. Dr. F. M. Wadley, Statistical Consultant for the then Bureau of Entomology and Plant Quarantine, U.S. Department of Agriculture, discussed the problem by correspondence as follows:

> Comparison of zones by areas shows that areas of two whole circles are in proportion to the squares of their radii. Doubling the radius gives four times the area, for example, but zones of different distances vary in area directly as the average radii. The areas of zones of given width have a linear relation to their average radii.

> Density may be considered as inversely related to distance with an hyperbolic relationship. Areal comparisons may be made but encounter the same difficulties as use of radii. These difficulties are those encountered as dispersing populations slow, come to rest and cease to spread outward. Such cessation is attributed to (1) increased space or area as distances increase and (2) loss (death or stopping) of part of the organisms. Visualization may be given to a dispersing population terminating movement at once. They would be one-half as abundant at 200 as at 100 feet. If the population spreads over the area as a whole abundance is inversely proportional to the square of the radius.

Two viewpoints were discussed by Worley (1939) regarding the most accurate method of interpreting growth data of fungal colonies in solid substrates. This is a form of dispersal. One viewpoint is that the use of radii gives a better criterion than the ratio of areas.

Various means or methods were considered by Gilmour *et al.* (1946) to measure and understand dispersal. One such method was the relationship of density or abundance to unit areas at distances. Radial distances, however, were eventually employed.

Comparisons of areal and radial units were made by studies on dispersal of leaf- and plant-hoppers through distance dispersal curves by Itô and Miyashita (1961).

Release of *Nephotettix cincticeps* by Itô and Miyashita (1961) in a paddy field was followed by recaptures on the third day afterward. Although other methods of computing relationships of dispersion were admitted two equations were utilized:

$$\text{log of no. of recaptured insects} = A - BS, \text{ and}$$
$$\text{log of no. of recaptured insects} = A - B'r^2, \text{ where}$$

S - r^2, or area, A, B, and B^1 are constants. Graphic means were used to indicate the relationships of recaptured leaf-hoppers to distance. (fig. 4).

Rapid rates of regression were shown to 40 m. Radical and areal relationships were reported by the authors to be "statistically significant for both cases."

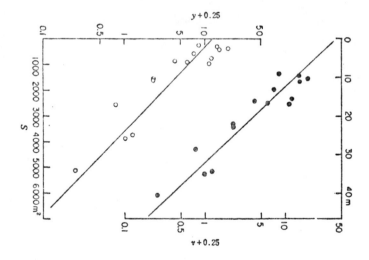

Fig. 4.   Regressions of recaptured leafhoppers. Ordinates show numbers of recaptured leafhoppers given in logarithms. Radial distances are shown in meters, at top; areal units in square meters, at bottom (reproduction of fig. 4 from Miyashita and Itô).

Point, band, block, or areal sources. Consideration may be given to the size and form of areas producing organisms. Different sizes of sources, or productive areas, are recognized. For instance, potato late blight disease spores, *Phytophthora infestans* (Mont.) deBy., may originate from a potato cull pile, from an eigthy-acre field, or from valley-wide areas of many acres. A cull pile would be recognized as a *point* source and a field or many adjoining fields as an *areal* source. A cage or box from which mosquitoes are released would be a point source; a marsh would be an areal source of mosquitoes. A row of potatoes could be a *line* source. The relative size of an area, in many or most instances, may be reduced to the mass or amount of organisms at the source. This is discussed in chapter 7. The time spent in producing or in emitting individuals may be short (instantaneous) or long (many days or weeks), both of which are relative terms that may have intermediate or varying occurrences.

Factors influencing dispersion. External and internal factors influence the dispersion of small organisms. These factors were divided by Williams (1923) into the "start," "course," and "finish" of the process of insect migration. His chart is presented in modified form here:

Table 1
Factors influencing dispersion

Dispersal Division	*Factors involved*	
	External	*Internal*
Start	Overcrowding, food shortage, wind, moisture (excess or deficiency), temperature, light (darkness), or the atmosphere	Hunger, sex impulse, periodicity of habit, imitation, maturity or development
Course	Aidants, barriers, temperature, wind, gravity, light (darkness), medium, moisture (wetness or dryness), density level, host density, attraction, or enemies	Species, stage (phase, generation or activity), sex, age, time, periodicity of habit, or hunger
Finish	Arrival at favorable or required conditions, failure of previous stimulus, barrier, light or (darkness) dark or enemies	Fatigue, end of reserve energy supply, development of sex organs or other physiological state

The degree of influence of these factors on distance of dispersal have been measured and recorded. Some factors are less subject to measurement or are open to conjecture. Each stage, phase, and activity, however, has its start, course, and finish.

Serial sampling: Data have been taken in the early parts of the dispersion journeys of a population of moving organisms and also at later intervals. These may be compared to determine differences as related to time. Graphic studies of curves drawn from data taken earlier tend to be steeper, with more curvature, and are higher on the graph than curves drawn from data taken later. Two sets of data pertaining to fungi and several to insects show this observation. Results of such dispersal behavior indicate that organisms early saturate the area near their source and tend later to extend the area of distribution.

Incidence of potato late blight disease was observed on two occasions, 24 days apart, and plotted by Waggoner (1952) to show sequential developments (fig. 5). Section A illustrates infections 18 days after the release of the sporangial suspension at the center of the plot. An unequal incidence of disease appeared northwest of the spore release center. Section B was made of the disease incidence 42 days after spore release, 24 days after the incidence ratings shown in Section A. Disease was found in all parts of the plot 42 days after spore release, in unequal amounts ranging from 2% to 20%. Incidences of disease in Section B appear inconsistently related in amounts to the "wind rose" of wind frequencies and directions in Section A. Winds were shown to have blown from all directions between 5 and 14 days after spore release, although in greater frequency from some directions than from others. Organisms causing the disease moved over the area, however, and inoculated leaflets in all part of the plot; then there were secondary or later cyclic origins of spores.

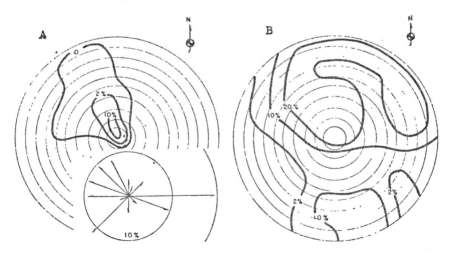

Fig. 5. Interval sampling of potato plot. Section A is plot 18 days after inoculum release at center of plot; at bottom is wind rose for days 5 to 14 after release. Section B is plot showing percentages of leaflets diseased 42 days after spore release in center of plot (fig. 1 from Waggoner).

Repeated, sequential, or serial observations contribute to a more complete understanding of a dispersing population than a simple set of observations. Patterns obtained during early and late sampling periods may differ considerably. All dispersion data pertain to some portions of the dispersal journey of a population and give information of that time. Organisms intercepted during a journey are precluded from completing a projected movement, but they have little or no significance to a mass of individuals in a journeying population. Incidences at the termination of the journey of a population might be expected to depict the final result of an expanded population.

Objectives of a study would determine to some extent the method or procedure of research employed. Several definitive examples of serial observations have been published, to which reference will be made below.

Most trees infected with the Dutch elm disease, *Ceratocystis ulmi Buis.* were within hundreds of feet of the apparent original source of the organism, as is shown by data taken by Liming *et al.* (1951) over a three-year period. Most diseased trees counted the second and third year were assumed to have originated from inoculations made originally. Regression curves in figure 6 show the relationship involved. Flatter curves are seen for 1945 and 1946 than for 1944. Generally low incidences of diseased trees were found at distances in excess of 1,000 feet each year of observation.

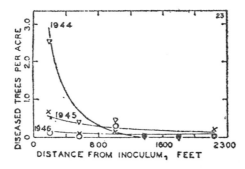

Fig. 6.   Dutch elm diseased trees at distances from the source on each of three years, showing successive reductions nearer the source on the 2nd and 3rd year of observations and slight increases in trees further from the inoculum source (data from Liming *et al.* [1951], curves from Wolfenbarger [1959]).

Peach trees in outer rows of orchards are infested with the plum curculio, *Conotrachelus nenuphar* (Herbst.), more abundantly than those in the middle. Curculios overwinter in woodlands adjoining the orchards and become more abundant in the outer trees and remain so

Dispersal Distances of Small Organisms

through the dispersing cycle. Primal and later curves are illustrated by data from Quaintance and Jenne (1912) in figure 7.

Curves were drawn to show the number of overwintered weevils shaken from trees in March, April, and May. More weevils were collected in March and April than in May. Egg deposition and mortality were factors that apparently reduced dispersal more distant than about eight rows and also reduced the total population.

Dispersal studies of overwintered plum curculios and brood larvae were found by Stearns and Haden (1932) and by Stearns *et al.* (1935) in outside rows of apple orchards. Overwintered adults and brood larvae were restricted largely to the nearer woodland hibernation quarters, while for the total season the populations tended to be less in row 1 and more in row 20 than those of the overwintered adults and 1st brood larvae (fig. 8).

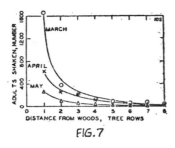

FIG. 7

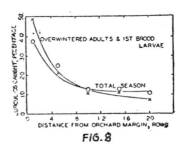

FIG. 8

Fig. 7. Plum curculio adults taken at distances from woodland overwinter quarters in March, in April, and in May, showing successive reductions in curculio populations with successive months (data from Quaintance and Jenne, curves from Wolfenbarger [1946]).

Fig. 8. Plum curculio dispersion into orchard, showing "overwintered adults" and "1st brood larvae" compared with "total season" (data from Stearns *et al.* and curves from Wolfenbarger [1946]).

A flatter curve appears for the total season's data than for the overwintered adults and first brood larvae, and the curve for the total season's data in a higher position at 20 rows' distance. This is suggestive of more distant dispersal by the adults from the first brood larvae than for the overwintered adults. It is a trend toward equalization of distribution in the orchard.

Collections of *Anopheles maculipennis* Meig. mosquitoes were made by Markovich (1941) in the Soviet Archangel province at different distances from known breeding areas on the Dvina River. Populations were determined in early summer, August and September, with relation-

ships illustrated in figure 9. Although the largest percentages of the
A. maculipennis were taken at the nearest distances each time of
determination, the percentages decreased each time. Shifts were made
in each of the determinations and at each distance point, and this resulted in flatter curves. Equalization of the species over the area with
time passage is thus indicated. Enroachment on more distant areas is
also suggested as a result, whether previously noninfested, lightly, or
heavily infested.

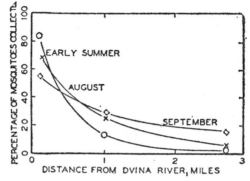

Fig. 9.   Incidence rates of *Anopheles maculipennis,* showing a tendency
toward equalization of populations by increasing flatness of curves from data
taken later (data from Markovich, curves from Wolfenbarger [1959]).

Dispersion of the wheat jointworm, *Harmolita tritici* Fitch, was
found by Chamberlin (1941) related to distance from wheat stubble.
Infestations were sampled by sweeping for adults early and by determining the percentage of larval infestations ultimately. Curves drawn
to show the regression for each sampling method are given in figure 10.

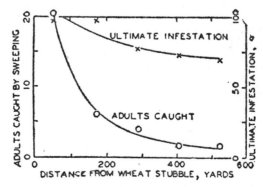

Fig. 10.   Dispersion of the wheat jointworm, "adults caught" compared
with "ultimate infestation" (data from Chamberlin, curves from Wolfenbarger
[1946]).

A curve of rapid descent for adults was reached at 522 yards but terminated at some unknown distance beyond. The ultimate infestation of near 60% at 522 yards was on a regression curve of slight inclination, which terminated at some distance considerably in excess of 522 yards. A wide difference between the initial and terminal sampling periods is indicated by these observations. Fair agreement of observed and curve values was found except for the ultimate infestations. 98% at each of the two nearest sampling distances.

Discontinuity in vertical dispersion. A discontinuity in the vertical gradients of the frit fly, *Oscinella frit L.,* was reported by Johnson, *et al.* (1962). Although the abundance of organisms ordinarily decreases with decreased elevation, these authors found an increase, termed "discontinuity," shown at 250 feet from pooled data, shown in table 2.

Table 2

**Discontinuity of frit fly decreases
by elevation from 9 to 1000 feet.**

Elevation, ft.	9	21	56	250	1000
Flies caught, no.	271	91	39	233	63

At 56 feet the lowest number of flies was found. This was a break or step called *discontinuity* by the authors.

Studies on vertical discontinuities were also made by Calnaido *et al.* (1965). Elevations from 0.016 to 32 m. were used to determine abundances during May, June, July, August, and September over grass shown in table 3.

Table 3

**Discontinuity of frit fly decreases
by elevations from 0.016 to 32.000m.**

Elevations, m.	.016	.065	.15	.30	.53	1.2
Flies caught						
Male	6124	1011	336	107	94	101
Female	4044	1138	385	135	153	124
Totals	10168	2149	721	242	247	225

Elevations, m.	2.4	5.0	8.4	14.9	18.1	24.0	32.0
Flies caught							
Male	543	308	148	64	46	27	27
Female	683	373	180	71	63	49	48
Totals	1226	681	328	135	109	76	75

Most frit flies were taken, as is usual, at the lowest collecting level; then decreases began with successively higher elevation, until at the 2.4 m. level the catch increased almost six-fold over the 1.2 m. level.

Decreases began again and continued to 32.0 m. the highest level at which collections were made. Other data also showed the discontinuity, a step in the gradient profile. Calnaido *et al.* (1965) reported that it "is peculiar to the frit fly; it occurs with both male and female but not with other insects." It occurred with the tiller and panicle generations. It seems likely that some factor, currently not understood, may have induced stratification in a number of instances.

SEGMENTAL PARTS OF DISPERSION AND INCIDENCE CURVES

Curves depicting the dispersal of small organisms tend to manifest certain characteristics. They have a starting point at the origin of the organisms, at zero distance. Most dispersal gradients are very rapid at the outset and begin to slope, then to flatten. That part of the curve which is most rapid and beginning to slope, perhaps ⅓ of the length, may be termed the *initial segment* (fig. 11). The part of the

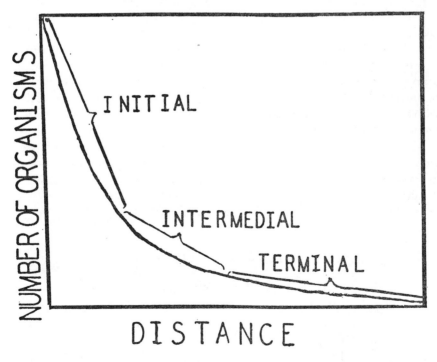

Fig. 11.   Parts of a graph depicting dispersion of organisms. Horizontal (abscissa) is distance, the independent factor. Vertical (ordinate) is number of organisms, the dependent factor. The rapid or beginning descent of the curve (gradient) is the initial; the hollow, or most curved part, is the inter-medial; and the more flattened portion is the terminal segment.

curve which is hollowed or most concave may be termed the *inter-medial segment*. As the distance increases the most concave portion becomes an ever more flattened, less curved part, which may be called the *terminal segment*. The terminal segment tends to approach but to remain slightly above the zero quantity. It may comprise almost one-half of the curve. Although the initial segment is more definite, especially the beginning, the other segments are more relative, without sharp points of demarkation. The final terminus of most individuals in a dispersed population will never be known.

Empirical randomness. There are instances of apparent lack of response to distance effects in many dispersion and incidence data. This applies more especially to the terminal segment of the dispersal range. The author has observed such phenomena on a number of occasions. It was observed with the potato flea beetle injuries to tubers (Wolfenbarger, 1940), European elm bark beetle twig crotch injuries, (Wadley and Wolfenbarger, 1944), incidence of serpentine leaf mines (Wolfenbarger, 1948) and the incidence of papaya viruses (Wolfenbarger, 1966).

In these studies, as is the usual practice, data were taken at radii from the source of the dispersing organisms until a static level of organisms or injuries was observed and there seemed little likelihood of much further change, except at some unknown distance. Fluctuations from zero to near zero or low numbers were found without apparently significant distance relationships. A state of distribution is evident that, without the application of statistical analysis, may be termed *empirical randomness*. An example follows.

Of 131 in a cherry orchard, increasing numbers of infected trees were shown by Demski and Boyle (1968) scattered over the orchard in an unordered fashion each year of the observations. No marginal nor directional influence appeared. Numbers of infected trees were counted in the figures given and are tabulated in table 4.

Table 4

**Number of sour cherry trees found in successive years
in a 131 tree sour cherry orchard**
(Data from Demski and Boyle 1968)

Year of observation	Number of trees infected
1955	3
1959	4
1961	7
1962	15
1963	18
1964	32
1965	86
1966	113

A low incidence of virus spread is indicated for eight years, until 1963, when eighteen trees (13.8%) were infected; then rapid spread occurred. Although the majority of infestations reportedly bordered previously infected trees, virus appeared without following a distance pattern response. This may be a random effect, as shown in figure 1 in the article by Demski and Boyle (1968), and suggests chance or random locations of diseased trees.

Discontinuity in horizontal dispersion. Discontinuity has not been recognized in horizontal dispersion as was observed for frit flies in vertical dispersion. Although variations were observed in horizontal movements, they were often or usually attributed to sampling procedures, barriers or other factors, biological or otherwise, rather than to discontinuity of gradients.

Half-density. An objective of some studies might include the distance to which some percentage, such as 2, 5, 50, 95, 98 (or some other) percent, of central density might be expected. Such an objective might include the distance for a minimum of pollination or a maximum of an insect pest or incidence of disease regarded as tolerable. It is recognized that the total dispersal of an organism may not be determined; nor might it be utilized if it were known. Estimated rates of organismal dispersion are obtainable, however, and are frequently employed for the interest and benefit of man.

Consideration of half-density (or other fractional density) and formulae for determining it at designated distances from the origin of organisms have been given by F. M. Wadley:

> Most studies on the dispersal of small organisms, fortunately for ease in the computations, are recorded in terms of density per unit rather than in terms of total population.
>
> Three descriptive regression curves have been tried and are listed (where Y is density and x is distance) as follows:
>
> $$\text{(1)} \quad \text{Log } Y = a + bx$$
> $$\text{(2)} \quad Y = a + b \log x$$
> $$\text{(3)} \quad Y = a + b \log x + c/x$$

Formula (1) gives a definite value for $x + 0$, the other formulae do not. Formulae (2) and (3) require the use of some value near $x = 0$, say $x = 1$ unit, for the density at the origin.

> Type problems, one from a simple and the other from a more complex regression, illustrate computations of half-density. A regression formula from data given by Bedard (1939) and calculated by Wolfenbarger (1945) serves as an example for a simple regression. In this example, percentages of trees were killed by the mountain pine beetle, *Dendroctonus monticolae* Hopk., at distances from newly constructed roads where logs and slash were left. The curve formula

$$Y, \% \text{ of trees killed}, = 63.3 - 53.3 \log x,$$
(distance from roadway in chains).

At what distance will only 5% of the trees be killed? The Y for log $x = 0$ (and $x = 1$) is the a value 63.3. This 63.3 is taken as the central value. Five percent of 63.3 is 3.2, which is set equal to Y as follows:

$3.2 = 63.3$ (log x). or
$- 60.1 = - 53.3$ (log x). We solve for
log $x = 60.1/53.3$ or 1.13, then

$x = 13.5$ chains, the distance estimated to have reduced the percentage of trees killed to 5% of the 62.3% killed at one chain. Three percent of the kill was calculated to be reached at 14.1 chains from the roadway.

Sometimes it may be desirable to arrive at some percentage or portional datum in the amount of incidence of organisms of a regression. Portional segments may be determined by two methods that were given the author in correspondence by F. M. Wadley:

Curves of the form
Log $Y = a + bX$

can be used to compute any desired percent. Care in manipulation of logs is necessary. Write the log of the desired percent as log Y and solve for X.

$Y = a + b$ log X,

can easily be solved for log X for a given Y. It is necessary to remember that log X is 0 at 1 unit distance, so that solution is approximate.
A curve of the form,

$Y = a + b$ log $X + c$ $(1/X)$

will present difficulty in solving for X except by approximations. It would seem best to carefully draw a graph of the curve and read off X's for desired Y's.

A more complex regression curve was computed by Wadley and Wolfenbarger (1944) on elm tree twig crotch wounds made by the smaller European elm bark beetle, *Scolytus multistriatus* Marsh., at distances from a pile of elm logs, a source of the beetles. The curve formula,

Y, % of crotches wounded, $= 27.94 - 8171$
(log x, distance from beetle source, ft.)
$+ 1276/x$, distance from beetle source, ft.

Fifty feet from the source may be taken as the starting point; its log is 1.70. Solving the equation,

$Y = 27.94 - (8.71) + 1276/50$
$= 27.94 - 14.81 + 25.52$
$= 38.65$

Fifty percent of 38.65 is 19.33, half-density. We try $x = 100$ and calculate the percentage injured,

$$Y = 27.94 - (8.71) (2.00) + 1276/100$$
$$= 27.94 + 1742 + 1276$$
$$= 23.28, \text{ which is more than half-density}$$
of the crotches injured at the 50-foot distance. Let us try $x = 200$,

$$Y = 27.94 - (8.71) (2.30) + 1276/200$$
$$= 27.94 - 20.04 + 6.38$$
$$= 14.28, \text{ which is less than half-density.}$$

Hence, the 50% (or half-density) of crotches injured is reached somewhere between 100 and 200 feet from the elm log pile, the source of the beetles. Other x-values between 100 and 200 feet may be tried to approach the half-density. A convenient way, however, is to plot the regression curve on cross-section paper and then read the x, or distance, value for the 50% or half-density of 38.65.

Dispersal and distribution. It is a presumption of parasite or predator release work that members of the released species will disperse. Distances to which they disperse may affect the effectiveness of a given organism. Dispersal distance data of *Lydella stabulans grisescens* R. D., a parasite introduced for control of the European corn borer, were given by Baker *et al.* (1949) and MacCreary and Rice (1949).

Introduction and release of the parasite in 1927 was followed in 1932 through 1938 by studies of parasitized corn borer larvae by Baker *et al.* (1949). Data obtained are reproduced in table 5.

Table 5

Incidences of European corn borer parasitization within radii of release site

Year	Parasitization, within radius of release site (miles)	
	3.5	7.5
1932	0.3	—
1933	2.8	—
1934	6.3	—
1935	7.6	4.4
1936	10.0	7.0
1937	17.1	9.6
1938	20.9	13.0

Successive increases in parasitization were found within 3.5 miles of the release site from 1932 through 1938. Successive increases were also found within 7.5 miles of the release site from 1935 to 1938. Less parasitization was found at the 7.5 than at the 3.5 mile distance. Since there were increases at each distance class each year, these might be expected to continue. An equalization might be expected, however, in the nearer, 3.5 mile, distance zone and to be observed later in the more distant, 7.5 mile, area.

Tachinid parasites, *Lydella stabulens grisescens* R. D., were released in 1941 and became established in corn borer larvae. Records of parasitization obtained by MacCreary and Rice (1949) were given for a five-year period, from ½ to 6 miles from the release site. These data are given in table 6.

Table 6
Annual incidence of European corn borer parasitization at miles from the release site

Year of observation	Miles from release site			
	0.5	2.0	4.0	6.0
1943	27	13	10	0
1944	25	16	9	11
1945	17	16	13	11
1946	33	28	27	29
1947	22	19	18	14

Parasitization was highest at the 0.5 mile distance and tended to decrease with longer distances each year of the observations. There was a tendency toward equalization of parasitization over the 6-mile (distance) range in 1945, 1946, and 1947. Parasitization, however, extended to distances in excess of 6 miles. Data on secondary, tertiary, and later cycles of parasitization involved are not given in the tabulation. Part of the data in the above three tables, therefore, may be rightfully considered as distributional rather than dispersal.

A futher indication of the dispersal of *Lydella stabulans grisescens* was given by Baker *et al.* (1949) around a release site seven years afterward. These data are given in table 7.

Table 7
Parasitization of European corn borer at distances 11 years after release of *Lydella stabulans grisescens*

Miles from release site	Percentage of borers parasitized
1	24.6
2	22.4
3	16.6
5	2.7
7	6.1

Distances in excess of three miles were shown to have reduced parasitization of corn borer larvae. How many cycles of parasites may have occurred within one, two, or seven miles of the release site is unknown. Intensification of parasitization is evident to seven miles.

I
INORGANIC FACTORS

Inorganic factors are comprised of inorganic constituents, originate outside the organism and affect active and passive disperser species. The inorganic factors are:

> Localities
> Directional (compass)
> Gravitational
> Medium
> Climatic
> Aidants
> Barriers

Although aidants and barriers are usually considered to consist of inorganic constituents, some possess organic components. Plant species that are host to some organisms may be restricted to certain definite areas. The presence or absence of such plants acts as aidants or barriers and restricts species to areas where the plant is present. Although such restriction could be an organic barrier, discussed below under the heading of Inorganic Factors.

2

Localities

Combinations of biological, chemical, and physical variables, such as available food supply, enemies, protection, soils, salinity, temperature, moisture, light, and other factors applicable to the life of specific organisms, characterize different localities. Localities may comprise very small areas or consist of large areas and cover many square miles. Often they lack definite boundaries. More complete understanding of the term *locality* may be obtained by studying such publications on ecology as the following: Allee *et al.* (1949), Andrewartha and Birch (1954). Cain (1944), Clark and Evans (1954), Clements and Shelford (1939), Oosting (1948), Coulter *et al.* (1931), Darlington (1957), Elton (1949), Heape (1931), and Odum (1959).

Because of the various factors having variable magnitudes in diverse localities, different rates of dispersion of organisms might be expected. Statistical analyses of various biological phenomena indicate that widespread differences frequently exist among localities.

Passive disperser units. The isolation of fields for producing virus-free potatoes is a general practice for the maintenance of healthy tubers for planting stock. Total isolation from all infestations is impossible, but partial isolation is practicable. Low incidences of virus diseases are "rogued" from plantings. Low incidences are found in certain areas such as the northern parts of the United States, the Canadian provinces, and in northern Germany. Many insects of the family Aphididae are effective vectors of many viruses and are more abundant in certain localities, such as the southern parts of the United States and of middle Germany (Quedlinburg), than in others.

Locality differences in virus-infected potato plants were given by Neitzel and Müller (1959) and graphically illustrated. Approximate percentages of virus-infected plants were determined from the graph (fig. 3) at the two locations and are given in table 8.

Table 8

**Virus infections at row distances
from the border, for each of two localities**

Distance from border (rows)	1	2	3	4	5	6	7	8	9	10
Location—										
Quedlinburg	86	54	36	40	26	25	27	16	17	21
Gross-Lusewitz	17	1	1	3	0	3	0	3	4	0

Outer rows contained more infections in each location than rows farther from borders, but differences between locations are very marked.

Significant differences in the incidences of potato mosaic were observed by Murphy (1921). These differences were rather widely spread over distance ranges through five rows (probably spaced about three feet apart) and are given in table 9.

Table 9

**Percentages of potato mosaic on plants at distances
from diseased plants at different locations in Canada**

Locality	*Distance from diseased plants, rows*					
	Between two rows	*Next*	*2nd*	*3rd*	*4th*	*5th*
Charlottesburg, P. E. Is.	45.0	27.5	15.0	17.5	22.5	6.2
Kentville, N. S.	32.5	22.5	13.7	12.5	10.0	1.2
Nappon, N. S.	55.0	31.2	18.7	32.5	20.0	18.7
Fredericton, N. B.	25.0	10.0	12.5	12.5	17.5	16.2
Lennoxville, P. Q.	57.5	42.5	0	2.0	5.0	5.0
Brandon, Man.	13.0	11.8	10.9	15.5	8.5	9.6
Indian Head, Sask.	4.5	2.6	2.2	0	0	1.8

Locality effects of potato rugose mosaic inoculations were given by Broadbent and Gregory (1948). These are summarized as percentages of diseased plants for three distance classes in table 10.

Table 10
Percentages of potato rogose mosaic on plants at distances from diseased plants in Canada

Locality	Inches from infector plants		
	38.5	79.0	118.5
Askham Bryan	16.8	3.4	2.3
Baston Fen	9.2	3.6	1.4
Bretton	1.0	0.7	0.3
Cardiff	27.1	6.4	2.5
East Malling	1.8	0.2	1.4
Goole	6.3	1.8	1.3
Harper Adams	22.6	5.4	2.6
Postland	7.8	1.9	0
Reading	15.5	6.2	2.5
Rothamsted	17.8	3.7	3.4
Slough	6.8	2.3	3.4
Sutton Bonington	25.6	4.8	4.1

More potato leafroll virus is spread in some locations than in others. Such incidences appeared more pronounced with leafroll than with mosaic disease, as is indicated above. Data on leafroll incidence from Murphy and Worthley (1920) are given as percentages in table 11.

Table 11
Percentages of potato leafroll on plants at different locations in Canada

Locality	Distances from diseased plants, rows					
	Between two rows	Next	2nd	3rd	4th	5th
Charlottetown, P. W. Is.	15.0	21.0	2.5	2.5	0	0
Kentville, N. S.	35.0	12.5	6.5	1.5	2.5	1.2
Nappon, N. S.	20.0	16.2	0	0	0	0
Fredericton, N. B.	5.0	0	3.7	1.2	2.5	1.2
Lennoxville, P. Q.	17.5	17.5	0	0	0	0
Ottawa, Ont.	72.5	50.0	23.7	22.5	13.7	10.0
Thunder Bay, Ont.	0	0	0	0	0	0
Brandon, Man.	13.8	5.6	12.3	3.2	7.2	8.0
Indian Head, Sask.	5.3	1.4	1.4	0	0	0

Incidences of potato leafroll transmission varied widely with locality, according to Broadbent and Gregory (1948). This is shown by data summarized according to percentages of diseased plants for three distance classes in table 12.

Table 12

**Percentages of potato leafroll on plants
at different locations in Canada**

Locality	Inches from infector plants		
	38.5	79.0	118.5
Askham Bryan	20.6	7.1	8.0
Baston Fen	32.2	12.0	10.1
Bretton	5.8	4.4	1.3
Cardiff	10.9	5.3	2.3
East Malling	29.2	6.7	5.7
Goole	11.8	8.2	6.2
Harper Adams	13.6	7.8	1.8
Postland	38.7	21.7	18.5
Reading	7.4	11.3	5.2
Rothamsted	29.4	11.0	10.8
Seale-Hayne	4.8	0.5	1.0
Slough	9.4	3.7	5.4
Sutton Bonington	31.7	5.4	3.6

Incidence of sugar beet savoy disease was studied by Coons, *et al.*
(1958). They found distinct locality differences. Regression curves were
drawn from the data given and are presented in figure 12 to show the
differences between locations. Although varietal differences in amounts
of disease were found (fig. 12), the data at each location were pooled

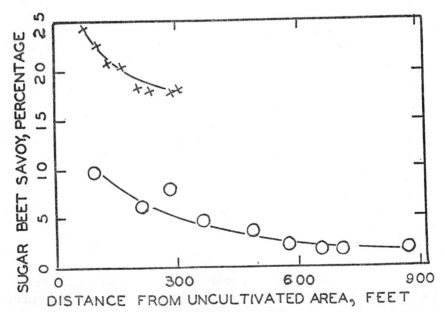

Fig. 12.   Incidences of sugar beet savoy disease. Upper curve—data from
different varieties grown near Pandora, Ohio (data from Coons *et al.*).

to show locality differences. It is of interest to observe that the regression constants were -8.6% and -9.4% at the Pandora, Ohio, and Mankato, Minnesota, plots, respectively. Similar curvilinearities, therefore, are indicated for the disease incidences at each location. Intensities of disease, however, differ. This may be attributed to localities or to one or more other factors.

Reforestation of burned over areas was observed near to and far distant from source trees surviving fires, according to Hofmann (1971). Dispersal of seed from source trees and seed remaining in the forest duff were recognized as origins of young trees following burns. More seeds sprouted and grew nearer seed that survived burns, however, than at remote distances from known source trees. Dispersion patterns characteristic for several species were shown as affected by distances and expressed as trees per acre. Locality differences are shown by two tree species, western hemlock, *Tsuga heterophylla* (Ref.) Sarge., and western red cedar, *Juniperus scopularum* Sarge., in table 13.

Table 13

Reforestation of western red cedar at two and four chain distances from seed sources in each of two locations

Tree species	Distance from seed source, chains	
	2	4
Table I. Knaiksu National Forest area		
Western hemlock	195	16*
Western red cedar	266	16*
Table V. Cowlitz area		
Western hemlock	1640	120
Western red cedar	2120	200

*"All areas beyond 3 chains" are termed 4 chains for these comparisons.

Locality effects on active and passive disperser organisms. There appears to be no data to show whether organisms that disperse actively are affected more or less than those that disperse passively. Some arguments suggest that energy from within an organism would be more constant than that from without and that actively dispersing organisms would be more variable. Other factors are operative, however, including control over movement by organisms that disperse actively. Locality differences of active disperser organisms may be greater than those of passive disperser organisms.

Spatial relationships of Polemoniaceae plants, *Linanthus parryae* (Gray) Greene, with blue and white flowers were studied by Epling and

Dobzhansky (1942). Blue colored flowers were found aggregated in a few areas and sporadically elsewhere, suggesting that through mutation or dispersal of pollen low frequencies occurred. Correlations were computed between the frequencies of plants with blue flowers in populations found at different distances. A graphic regression type curve was prepared to show the results (fig. 13). Consistent trends of changes were found with greater distances, as given in a straight-line relationship. Populations from nearby areas are more closely related than are those from more remote areas.

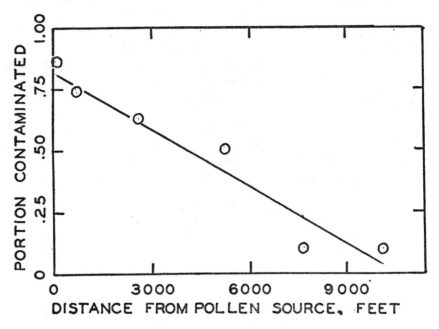

Fig. 13. Incidences of *Linanthus* contamination (data from Epling and Dobzhansky).

Active disperser organisms. Organisms that disperse actively are also affected by locality. Distances to which the boll weevil, *Anthonomous grandis* Boh., moves into woods adjoining cotton fields for hibernation is in terms of a few hundred feet. Localities may be expected to influence these distances owing to different factors, or perhaps to the relative magnitude of the different factors. Observations in two counties in South Carolina by Fye *et al.* (1959) show some wide differences. These are summarized, grouping the replications to give the average number of beetles per two square yards of area in table 14.

Table 14

Cotton boll weevil dispersal to distances
in each of two locations in South Carolina

Distance from cotton fields, (ft.)	15	45	75	105	135	165	195	255	285	315
Darlington Cnty.	8.1	12.9	6.4	7.5	7.1	4.9	1.2	0.6	—	—
Florence Cnty.	17.8	20.6	8.9	6.6	7.7	4.4	4.1	0.7	1.7	0.4

More beetles were found within 100 feet of cotton fields in Florence County than in Darlington County. At distances beyond 100 feet, however, similar rates of dispersion were observed. Local conditions, such as more and larger trees, might account for this difference.

Trees in outer orchard rows of peach, *Prunus persica* (L) Stokes, are more heavily infested by the plum curculio, *Conotrachelus nenuphar* (Herbst.), than are trees more distant from the border. Similarities were found by Snapp (1930) in different locations, as shown by the data in table 15.

Table 15

Plum curculio dispersal to distances
in each of two locations

Number of curculio found at 2 locations:	Distance from orchard margin rows							
	1	2	3	4	5	6	7	8
Author's Table 56:	358	257	198	155	123	96	74	54
Author's Table 57:	321	217	189	154	127	105	87	70

An examination of the data shows: (1) more curculios in rows 1-4 of table 56 than in table 57, (2) equal numbers near rows 4-5, and (3) more beetles in rows 5-8 of table 57 than in table 56.

Locality differences in incidences of twig crotch wounds made in elm trees by the smaller European elm bark beetle, *Scolytus multistriatus* Marsham, were reported to be rather marked (Wadley and Wolfenbarger, 1944). At least part of these wounds may be attributed to the fact that there was a larger population of beetles in the New Jersey area than that in the Connecticut area. Differences in localities were found to be much wider nearer the beetle origin than at more remote distances. This is shown by the percentages of crotches containing wounds, as computed by the regression formulae (Wadley and Wolfenbarger, 1944) in table 16.

Table 16

Incidence of twig injuries in elm trees at distances
from the beetles' source of
the smaller European elm bark beetle

Locality	Distance from beetle source (feet)	
	200	*500*
New York	23	4
Connecticut	44	3

Locality differences in numbers of material galleries made by the smaller European elm bark bettle were shown from reports by Wolfenbarger and Jones (1943) and Whitten (1938). Regression curves (Wolfenbarger, 1946) showed similar rates of incidence, although more galleries were reported by Wolfenbarger and Jones (1943) than by Whitten (1938).

Dispersal of native elm bark beetle, *Hylurgopinus rufipes* Eich., from its origin in logs or dead trees to living trees for hibernation was studied in each of two locations (Wolfenbarger, 1941). In the Allamuchy, New Jersey, area many bettles had emerged from chemically treated trees and in the Lanesboro, Pennsylvania, area smaller numbers of beetles had emerged from slash left from clearings for mine timbers. Dispersion occurred to distances in excess of 800 feet in the New Jersey area and to near 100 feet in the Pennsylvania area. More dense beetle populations in the New Jersey area may account for part of the widespread differences.

A common experience is that mosquitoes and other insects which seek meals of human blood are more abundant in some areas than in others. Man tends to avoid those areas in which he is discomforted by insect attacks. One of the many factors that affect the abundance of these pests, however, is the dispersion activity. Reference is given here to influences of locality in the dispersion activity.

In each of two locations dispersion of the dark and tan rice-field mosquitoes, *Psorophora confinnis* (L. A.) and *P. discolor* (Coq.), extended miles, according to Horsfall (1942). A regression curve drawn from the data taken at each location, (Wolfenbarger, 1946) shows that one curve may be practically superimposed on the other. Although one curve has more curvature than the other, it extends about one-half the distance of the other and is shown a different scale; the dispersion rate is nearly equal for the two locations.

Releases and recoveries of laboratory-reared houseflies were made in a city garbage dump where many flies were breeding and in a residen-

tial section of a city where flies were few (Schoff and Siverly, 1954a).
Regression curves drawn to represent the dispersion rate in each location
(Wolfenbarger, 1959) (fig. 14) were almost superimposed, with similar
positions and rates of slope. Similar rates of dispersion were shown,
therefore, regardless of locality.

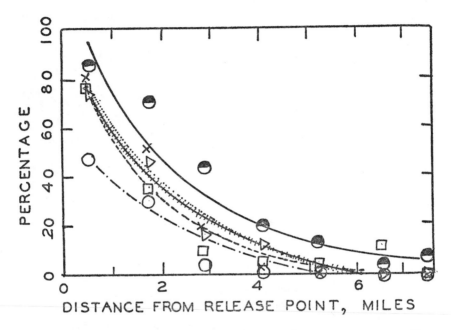

Fig. 14.   Dispersion of the housefly from different release sites. Upper
curve represents data from "hog farm." Lowest curve represents data from
"rendering plant." Barred curve represents data from "meat packing concern."
Dotted curve represents data from the "lettuce dump." Dash and dot curve
represents data from "poultry farm." (Data from Schoof and Siverly.)

Superparasitization of Japanese beetle larvae, *Popillia japonica* Newm.,
by hymenopterous insect, *Tiphia vernalis* Rhw., extended to distances
in excess of 90 feet from feeding areas (Gardner, 1938), (fig. 15). Similar
rates of superparasitization were found at Philmont and Rushland,
although slightly different levels of egg deposition were found.

Trees, hedges, and banks or ditches with a thick, low-growing
covering were regarded by Wright and Ashby (1946) as *"one of the
most important attributes of a field in determining the degree and position of the attack"* from the carrot fly, *Psilia rosae* F. Also see figure 2.
Effects of ground covering is shown in table 17 by mines per 100 carrots.

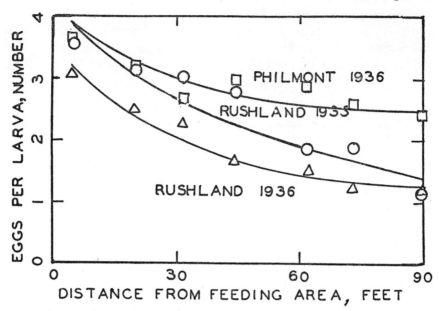

Fig. 15.   Effects of localities on superparasitization of Japanese beetle larvae by *Tiphia vernalis* (data from Gardner).

Table 17

**Incidence of carrot fly mines
in carrots in each of two localities**

Source (locality)	Distance from field margin (yards)					
	0	5	15	30	40	50
Curves I and II	69	53	28	14	21	29
Curve III	4	6	2	2	2	0

Invasion of the southeastern part of the United States by the cotton boll weevil, *Anthonomous grandis* Boh., began in 1892 and was accomplished in 1922, as indicated by figure 16, a map from Hunter and Coad (1923) showing the rate as spreading. Circular segmental patterns are indicated with tendencies to focus on the original site of the initial invasion. Such patterns may be the expected occurrence. Such an invasion indicates locality differences. Weevils simply entered localities where the host plant, cotton, was growing. Directional influences are indicated toward the east-northeast because (1) the Gulf of Mexico barred plant growth to the south and (2) climatic temperatures prevented cotton production to the north.

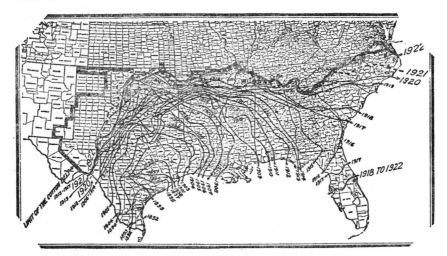

Fig. 16.   Spread of the cotton boll weevil through the cotton-growing areas of the southeastern United States (copy of fig. 1 from Hunter and Coad).

MEDIUM

Organisms must pass through a medium or media, regardless of how they disperse. Air, water, and soil are the common media through which organisms disperse. Various characteristics of the media may induce a change, such as hastening or retarding the dispersion process. It is difficult or impossible, however, to distinguish between medium and locality, and between medium and climatic factors. There appears to be no important data collected on dispersion that may be attributed to different media. This may be readily understood in view of locality differences and various climatic factors involved in environments. Successful dispersal movements are related to the total and varied conditions —syndromic effect—rather than to individual components.

Air is the most common and important medium through which small organisms disperse. Components of the air are required for the maintenance of life. Aerial travel is used by most organisms as a transient interlude between two local sites. All organisms are dependent on soil, water, and objects on the ground or on the water for most of their existence.

Water in one form or another is frequently an agent of dispersion. Rainfall during spore porduction periods was suggested by Wilson and Baker (1946a) as a factor reducing distances to which conidial spores of *Sclerotinia laxa* Ader and Ruh. were dispersed by wind. It is likely that rainfall during the dispersal of most organisms reduces the distances reached. On the other hand, the spores that are dispersed in rainy intervals may germinate and develop to maturity more rapidly than those dispersed in dry periods. Further considerations are given to water, moisture, and other climatic factors in chapter 5.

Dispersion of soil-inhabiting organisms is represented by very few definitive data. Less is probably known of the dispersion of soil-inhabiting bacteria, fungi, insects, and nematodes than of air-inhabiting organisms of the same groups.

Data on the incidence of damping-off disease caused by *Rhizoctonia solani* Kuhn were given by Blair (1943), who found differences in incidence rates for two soil types. These data are given as percentages of plants damped-off in table 18.

Table 18
Incidence of damping-off on two different soil types

Distance from source of inoculum (cm)	1	2	3	4	5	6	7	8	9
Soil type									
Harwood	100	45	8	31	7	0	16	5	0
Allottment	100	75	14	70	60	54	31	32	10

Higher percentages of damped-off plants are seen in the Allottment soil type than in the Harwood soil types, especially at distances in excess of 2 cm. Although soil types was undoubtedly of importance in affecting the differences in damping-off, water content or other factors also may have been important.

Take-all disease of wheat by fungus *Ophiobolus graminis* (Sacc.) Sac. spreads through soil by root contact, according to Wehrle and Ogilvie (1956). More rapid spread was reported in alkaline and loose-textured soil, although data were not given to show differences. The longest distance was given as about five feet.

The above references were those of passive disperser organisms as affected by media. Differences doubtless exist for active disperser groups but data were lacking.

Eelworm movement through soil was found to be affected by particle size by Wallace (1958a). Optimum size was 150-250 microns for *Heterodera schactii* Schmidt and 250-500 for *Ditylenchus dipsaci* Filipjev. Optimum temperature for dispersal of both species was given as 15°C.

Climate and dispersal were discussed by Greenback (1956) as factors in the initiation of outbreaks of the spruce budworms, *Choristoneura fumiferana* (Clem.). These outbreaks have been traced for nearly

200 years and are generally attributed to budworms generally believed to originate in an area in which differences are implied or inferred from other localities. Dispersal from specified localities is therefore a mode of development of destructive infestations. Climatic variations and many other factors, present in great complexes, were affecting the dynamics of epidemic spruce budworm populations discussed under the editorship of Morris (1963).

Discussion. Locality differences are evident in the above examples. Density of organisms at the source and meteorological or other factors may account for some differences in incidence or dispersion curves. Populations of curves, incidence of white pine blister rust, and dispersal of the housefly indicate similarities and dissimilarities in the regression curves determined from different investigators, localities, and whatever other factors were involved (fig. 17). Eight curves show the results of

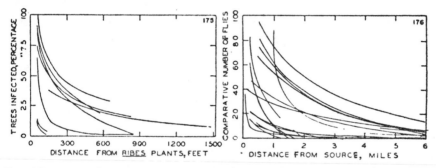

Fig. 17. Incidence of white pine blister rust (left) and dispersion of house-flies (right). Curves were drawn by Wolfenbarger (1959) from data given by various authors.

white pine blister rust infected trees in terms of hundreds of feet. Seventeen curves show comparisons of dispersion of the housefly under different conditions, but that distance range is in terms of miles. Curves initially high have a tendency to remain high with increased distances and to terminate at more remote distances than curves where the initial position is low. Other factors may be confused with locality, although the present evidence suggests that localities have little effect over the distance dispersal of a species.

3

Directional, or
Polar, Influences

Dispersion implies a planar horizontal movement in one or more directions from the source. Organisms that move in all directions from the origin disperse omnidirectionally. Such dispersion is fundamental and implies (1) a lack of stimuli in one or more directions or (2) stimuli of equal value surrounding the organisms. Organisms that move in one direction from the origin disperse unidirectionally. Such dispersion is in response to stimuli of unequal magnitude in one direction, or from all directions except one, or, perhaps, in response to a particular stimulus. Unidirectional dispersion is the more specialized movement. *Migration,* one feature of which is unidirectional movement, may be considered a specialized form of dispersion. Not all unidirectional dispersion, however, is termed migration.

Organisms often move more in some directions than in others. Several factors may affect directional dispersion. *Air currents* are probably one of the most important factors. This is illustrated in patterns shown by Haskell and Dow (1951) (fig. 18). Mosquito prevalences found in studies by Gillies (1961) are shown in figure 19. *Wind,* a climatic factor, is a separate heading discussed in chapter 5. Ground covering, host densities, topographic features, temperatures, population densities, shade and other factors also affect directional movement. Although winds are predominantly from one direction, there are usually wind changes in most locations. Such changes are often sufficient to scatter organisms omnidirectly some time during the dispersion period of species. This may be seen in figure 18 or 19 when actually the directional differences may be more apparent than real. Uniformity of movement about the origin is not a requisite for omnidirectional dispersion.

Omni- and unidirectional dispersion. Differentiation between omni- and unidirectional dispersion of a species may be difficult and at times impossible. Available evidence is sometimes indicative of the presence or absence of direction influences. Sometimes there is evidence for naming the factor affecting directional movement. Sometimes differences between directions are evident but factor or factors responsible are

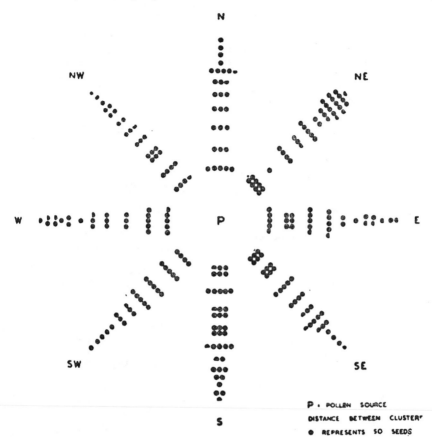

Fig. 18.   Directional relationships between seed-setting and distances from pollen source along stringers of detasselled sweet corn plants (fig. 1 from Haskell and Dow).

not known. Also, in some instances differences may be expected but not observed.

The attempt is made here to present instances of omni- and unidirectional dispersion. It is recognized in doing this that (1) some instances may be listed under one heading that belong to the other; (2) some instances should be given but are presented in another chapter; and (3) some instances are given under directional influences that belong in another chapter. It is conceivable that one or more stages or activities of a species would disperse unidirectionally and that the others would move omnidirectionally.

Tower (1906) referred to "natural highways which afford lines

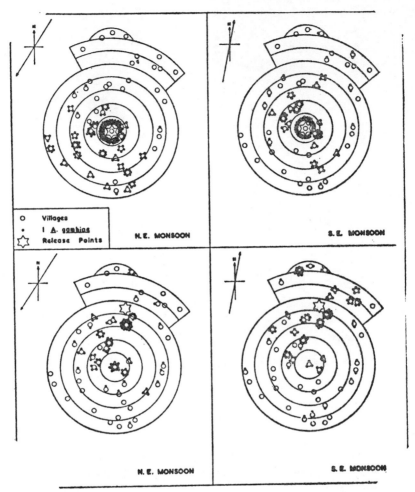

Fig. 19.   Dispersion of marked *Anopheles gambiae* mosquitoes in relation to prevailing winds (fig. 6 from Gillies).

of least resistance to migration." Directive factors preventing equal unidirectional dispersion were recognized but unnamed.

Unidirectional. General unidirectional movements by successive generations are recognized for many species. A brief discussion of this is given, although this subject is essentially in the field of distribution of a species and not of dispersion. Through invasion of a land new to a species its distribution may proceed in one general direction from the point of initial colonization. Examples of westward movement of species in the United States include the following: chestnut blight, *Endothia parasitica*

(Muir.) A. & A. A.; white pine blister rust, *Cronartium ribicola* F. von Wald.; Dutch elm disease *Ceratocystis ulmi* Buis.; cabbage butterfly, *Pieris rapae* (L.), given in the classical report by Scudder (1887); European corn borer, *Ostrinia nubilalis* (Hbn.); and Japanese beetle, *Popillia japonica* (Newm.). Since these organisms were introduced on the eastern coast in the eastern part of the United States and since favorable media for growth and reproduction extended westward from the point of introduction, a unidirectional movement might be expected. Successive cycles of dispersion that continue long enough in one direction, however, would eventually place a species in an area or region where, owing to climatic and other factors, conditions would be less than optimum, and further movement in that direction would result in losses to the species.

Influences of *polarity*, or of the heavenly bodies, are not generally recognized in the dispersion of organisms. Specific recognition was given by Lindroth (1953), however, to the "influence of the sun on the flight direction" of *Oodes gracilis* Villa in its westward movement.

Initiation and termination of flight directions of *Aelis integriceps* were related by Brown (1965) against, with, and across the sun's rays. Data from this study are given in table 19.

Table 19

Response of *Aelis integriceps* to sun's rays

Flight		Direction in relation to the sun		
		Against	With	Across
Initiation	All records (No.)	558	222	299
	All records %	51.7	20.6	27.7
	Wind over			
	6 ft./sec., No.	163	27	31
	6 ft./sec., %	73.8	12.2	14.4
Termination	All records (No.)	501	138	448
	All records %	46.1	12.7	41.2
	Wind over			
	6 ft./sec., No.	188	4	29
	6 ft./sec., %	85.1	1.8	13.1

Most flights were against the direction of the sun at initiation and at termination. More flights were across than with the direction of the sun. Record percentages of 73.8 and 85.1 were observed at wind speeds of over 6 mpr. against the sun's rays. Tabulation of observations pertaining to wind are given in chapter 5.

A discussion on the migration of the monarch butterfly, *Danaus plexippus* L., in North America was presented by Urquhart (1966). This insect is recognized as one that moves northeastward from its overwintering areas along the Gulf of Mexico. Eggs are deposited on host plants along the way to produce the first generation, which moves northward and in its turn lays eggs to the northern limits of its distribution. Southerly flights are believed to begin in July, growing stronger during August and September, and moving to the southwest into Mexico for the winter. Directional responses were shown for movements without distance data.

Unidirectional movements suggest that some stimulus was operational, such as might be provided by the earth's magnetic field. A lack of response to magnetic fields, however, was shown by Williams *et al.* (1942) in controlled cage tests. Locust nymphs, *Schistocerca gregaria* (Forsk.), adults of *Calandra granaria* (L.) and *Araecerus fasciculatus* (Deg.), and larvae and adults of *Tenebrio molitor* L. were exposed to electromagnetic fields. Positioning of the insects indicated that the insects made no response to the magnetic fields. This confirms observations of free-moving organisms, in which omnidirectional dispersion is usually observed. Omnidirectional dispersion is discussed below.

Omnidirectional dispersion. Detailed studies on the incidence of loose smut of wheat, *Ustilago tritici* Pers., were reported by Öort (1940) for a two-year period. The number of smutted plant heads observed per 100 m² at selected distances ranging from 14.7 to 15.2 m. from the source for each of the four cardinal directions is given in table 20.

Table 20
Unidirectional incidence of loose smut of wheat

Direction	North	East	South	West
Smutted heads (No.)	7.4	36.3	3.7	4.2

(interpolated)

Although prevailing winds spread more spores toward the east, it was concluded that the organism spread nearly equally in all directions over the experimental plots to distances of 100 m.

Epidemiological studies of wheat stem rust, *Puccinia graminis tritici* Eriks. & E. Henn., by Schmitt *et al.* (1959) indicated a "roughly elliptical pattern of outmovement." This indication was presented by diagrams showing the progress of rust spread from infection foci (fig. 20). Such spread, by a passive disperser organism, may occur more frequently than is often recognized. Rust developed in all portions of the field regardless of wind direction. Disease developed earliest, however, toward the southwest, in response to the prevailing wind movement of the preceding two weeks.

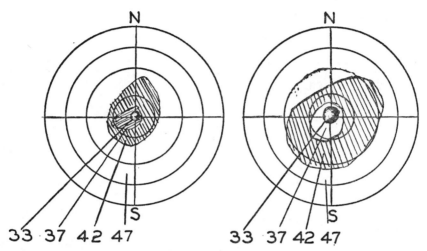

Fig. 20. Time required for stem rust of wheat spread from the 90-pustule concentrated focus to a 0.5 pustule per culin level. Numbers represent days since infected plants were placed in the field (from figs. 2 and 3 from Schmitt *et al.*).

White pine blister rust infections of *Cronartium ribicola* F. von Wald were reported by Buchanan and Kimmey (1938) as "comparable in all directions from the central ribes." Although skewness was observed, it was attributable to hill slope, or gravitational influences. A general reduction in number of cankers per million pine leaves, *Pinus monticola* Dougl., at distances from *Ribes* (taken from author's graph) is shown in table 21.

Table 21

Omnidirectional dispersal of white pine blister rust

Distances from *Ribes*, feet (mid-class points)	5	15	25	35	45	55	65	75
Cankers, per million leaves	45	22	8	4	2	1	1	1

Incidence of white pine trees infected with blister rust around black currant bushes, *Ribes floridum* L., reported by Posey and Ford (1924), was greatest in southern and least in eastern directions. Similarity of data taken from west and north of black currant bushes justified a combination of data from the two directions. These are summarized in table 22.

Table 22

Directional dispersal of white pine blister rust

Direction	Distance from the black currants (feet)						Total
	50	*150*	*250*	*350*	*450*	*550*	
East	65	15	4	2	3	3	92
South	100	63	58	45	18	9	293
West & North	86	55	37	25	17	10	230

These data show omnidirectional, although unequal, movement of spores from the source, with the lowest incidence eastward.

Cross pollination in onions, *Allium* sp., was found by Erikson and Gabelmen (1956) to have been unaffected by directional influences. Yield data from male-sterile plants surrounding pollen-fertile plants were homogeneous as to directions. Yield data from the 18-inch spacing, Plot B, 1952, are used to illustrate the rate of cross pollination and are given in table 23.

Table 23

Omnidirectional dispersal of onion pollen

Distance (ft.)	*1*	*4*	*7*	*10*	*13*	*16*	*19*	*22*	*25*
Yields (seeds/head)	563	439	389	357	334	315	300	287	275

Considerable attention was given to natural crossing in cotton, *Gossypium* sp., by Afzal and Khan (1950). Insects are the principal agents dispersing cotton pollen. They pollinate cotton flowers according to their instinctive behavior and in these observations moved pollen almost equally in all directions. Data taken over a two-year period and converted to percentages for each of the eight directions are summarized in table 24.

Table 24

Omnidirectional dispersal of cotton pollen

Direction

Year	North	East	South	West	N.E.	S.E.	S.W.	N.W.
1945-46	0.0428	0.0114	0.0253	0.0401	0.0498	0.0562	0.0573	0.0517
1946-47	0.0441	0.0242	0.0439	0.0520	0.0484	0.0493	0.0531	0.0523

Pollination of sweet corn, *Zea mays* L., with wind as the agent of pollen dispersion was studied by Haskell and Dow (1951). Pollination was found to vary significantly by distances and insignificantly by directions. A combination of data from all directions was, therefore, justifiably made to show the overall regression rate and is given in table 25.

Table 25
Omnidirectional corn pollen dispersal

Distance from pollen source (ft.)	36.3	28.1	23.4	20.0	17.4	15.2	13.9	11.8	10.4	9.2	8.1
Mean seed set (%)	4	8	12	16	20	24	28	32	36	40	44

Data on directional influences of the common periwinkle, *Littorina littorea* (L.), show but little effect (Smith and Newell, 1955). These data are summarized in table 26.

Table 26
Directional periwinkle dispersal

Direction of movement	North	East	South	West
Number of periwinkles	109	63	80	73

A popular conception of grasshopper dispersal is of swarms moving across vast areas in one direction. Parker, *et al.* (1955) recorded 481 detailed observations on swarms in 1939 and 132 records in 1940. These flights were classified as to direction of movement, including milling about without directional stimulus, and were given in terms of percentages as shown in table 27.

Table 27
Directional grasshopper dispersal

					Direction				
Year	N	NE	E	SE	So	SW	W	NW	Without direction
1939	6	10	7	12	4	8	20	21	11
1940	6	11	13	15	12	8	6	16	14

Most flights were observed moving toward the northwest, while the southeast direction ranked second. Obviously, grasshoppers flew in all directions. Many flights were milling about without direction. Flights were usually associated with low-level prevailing wind.

A generally accepted view of aphid activities is that aphids remain on the plants at wind speeds in excess of 4-5 miles per hour, Johnson (1954). This suggests that aphids are motivated to retain control of their movements. Greenslade (1941) reported that strawberry aphids, *Capitophorus fragariae* Theob., "fly in all directions and cannot be much affected by the direction of the wind since they do not fly unless the wind speed is very low."

Emphasis was given to wind direction and infestation of bean fields of *Aphis fabae* Scop. in studies by Taylor and Johnson (1954). Margins of fields were often infested more heavily than internal areas. Data from four sets of figures are summarized as mean number of colonies three inches and over in the final count in table 28.

Table 28

Directional bean aphid dispersal

	Northerly	Easterly	Southerly	Westerly
Mean no. colonies	26.9	26.4	52.3	34.6

Intensity of aphid infestations was associated with shelter at field margins and with wind direction, especially during the initial arrivals of dispersing aphids.

Localized dispersal of the six-spotted leafhopper, *Macrosteles divisus* (Uhl.), was studied by Linn (1940) using dyed insects. Data from two experiments are summarized in table 29 as percentages of the total number of dyed individuals collected in different directions.

Table 29

Directional dispersal of the six-spotted leafhopper

Stained M. divisus over crop areas	N	NE	E	SE	S	SW	W	NW
Diversified crop plants	3	6	13	16	35	23	0	3
Uniform (timothy) plants	11	19	27	17	9	6	4	7

Wind may have influenced direction of movement toward the south, although an omnidirectional trend is observed. Any effect of crop plant in influencing direction is unknown.

The effect of direction on dispersal of the cotton boll weevil, *Anthonomous grandis* Boh., was reported by Fenton and Dunnam (1928). Screens covered with a "sticky" material were placed in fields to catch and hold weevils contacting the covering. Weevils caught in the adhesive were counted, and the data are summarized in table 30.

Table 30

Directional dispersal of the cotton boll weevil

Direction	North	East	South	West
Weevils caught (No.)	202	221	190	217

Weevils were not blown by the wind, they did not drift with it, nor were they found flying against it, according to the authors. This is another instance in which the insect is shown to maintain control over its flights.

Wind was believed by van Zwaluwenburg and Rosa (1940) to be the most important factor affecting directional dispersion of the sugar cane beetle borer, *Rhabdocnemus obscura* Bsd., although there were no significant differences among directions, according to the author. Upwind and downwind movements, determined for each sex, were measured in feet and in days from the release point. These data are presented by means and standard errors in table 31.

Table 31

Compass directional and wind directional
dispersal
of the sugar cane beetle borer

	Mean distance (feet)		Mean time (days)	
Direction	Male	Female	Male	Female
NNE (upwind)	370 ± 20	360 ± 10	55 ± 5	42 ± 3
SSW (downwind)	465 ± 21	447 ± 28	38 ± 4	30 ± 3

Extensive data on the dispersion of the primary screwworm, *Cochliomyia macellaria* (F.), the fleeceworm, *Phormia regina* (Meig.), and the housefly, *Musca domestica* (L.), were given by Bishopp and Laake (1921). The recovery of marked and liberated flies was recorded for each fly by distance and direction from the release point. Data were recorded for *C. macellaria* and *M. domestica* for each of three distance range groups and for *P. regina* in two range groups. The data for *C. macellaria* are summarized by distance classes and directions in table 32.

Table 32

Directional dispersal of a screwworm, by distance classes

Distance range classes (miles)		Direction			
		North	East	South	West
0.4 - 3.5		388	111	92	364
4.1 - 8.2		72	55	63	81
7.0 - 17.8		0	2	2	4
	Totals	460	168	157	449

Most flies were taken north and west of the release point; fewest, south and east.

Data pertinent to *Phormia regina* are given in table 33.

Table 33

Directional dispersal of a fleeceworm, by distance classes

Distance range classes, (miles)		Direction			
		North	East	South	West
0.4 - 3.5		15	4	5	15
7.0 - 17.8		73	55	63	81
	Totals	88	59	68	96

Most flies were taken north and west of the release point; fewest, south and east.

Data pertaining to the housefly, *Musca domestica,* are given in table 34.

Table 34

Directional dispersal of the housefly, by distance classes

Distance range classes (miles)		Direction			
		North	East	South	West
0.4 - 3.5		1,325	1,059	321	889
4.1 - 8.2		33	8	19	9
7.0 - 17.8		1	1	10	1
	Totals	1,359	1,068	350	899

Most houseflies were taken north and east of the release point; fewest, south and west.

These tables show a trend for more flies of each species taken north and west of the release point and fewest, south and east. This trend is evident in both near and more distant range classes. More flies, except for *Phormia regina*, were taken in the 0.4 to 3.5 mile distance range classes than in each of the more distant range classes. Although unequal numbers of flies were recorded in the different directions, omnidirectional movements are shown for the three species over all of the distance range classes.

A detailed study of *Drosophila funebris* (F.) and *D. melanogaster* Meig. fly dispersion was given by Timofeef-Ressovsky (1940a, 1940b). The procedure involved in these studies was the release of marked flies and recovery by trapping of a portion of those released. *D. funebris* dispersed practically equally in all directions, according to the author's figure 3. *D. melanogaster* were slightly displaced directionally from the release point (author's fig. 4). That displacement was attributed to the wind.

Studies of dispersion rates were made of *Drosophila pseudoobscura* Frowola flies by Dobzhansky and Wright (1943). In four experimental releases of orange colored flies among a wild population of flies, it was found that the number of released flies collected in the different directions varied significantly. Wild flies also were collected in significantly larger numbers in some directions from the release point than in others. Although most flies were collected in the western direction and fewest in the northern direction, omnidirectional movements are shown in table 35.

Table 35
**Directional dispersal of a drosophilid fly
by distance and day after release**

Day after release		North	East	South	West	Totals
				Direction		
1		309	375	373	394	1,449
2		195	384	345	421	1,345
3		126	196	267	273	862
4		102	114	170	197	583
5		106	67	70	104	347
	Totals	836	1,136	1,225	1,389	4,586

Directional influences, according to the authors, were never large enough to produce striking displacements of the flies. More flies were collected near old trees, in dense vegetation, and in moister locations, indicating nonuniform terrains of the experimental areas. It was concluded that favorable sites for attracting flies were more responsible for fly abundance in some directions than were purely directional influences.

Spatial and directional effects of an introduced parasite, *Anastatus bifasciatus* Fonsc., of the gypsy moth, *Porthetria dispar* (L.), were ob-

Dispersal Distances of Small Organisms

served by Crossman (1917). Data taken in the eastern and western directions from the release point were so similar as to justify combination for computing a regression curve by Wolfenbarger (1946). Separate regression curves were drawn from data taken from southern and northern directions. More parasitization was found in eastern and western directions and lowest in the northern direction. The second lowest total was in the southern direction. Observed values in terms of percentage indicate the incidence of parasitization in table 36.

Table 36

Dispersal of an introduced parasite of the gypsy moth larvae by distance and direction

Direction from the release point	*Distance from the release point (feet)*							
	100	*300*	*500*	*900*	*1,500*	*3,600*	*4,200*	*5,100*
North	26	7	1	7	25	9	0	0
Avg. E. & W. combined	35	28	38	30	29	15	15	5
South	26	20	26	22	28	29	5	0

Similar rates of decrease were found by Wolfenbarger (1946) in regression coefficients of the formulae computed from the data in northern and southern directions -11.67 and -11.71 respectively. A slightly more rapid rate of decrease in egg parasitization was found in the coefficient (-15.60) computed by combining data from the eastern and western directions.

Data on prevalence of honeybees, *Apis mellifera* L., in different directions and at different distances from apiaries were given by Eckert (1933). Data from the northeastern, southeastern and southwestern quadrants were so similar that they were justifiably combined for the computation of a single regression curve. Data from the northwestern quadrant were different, and a separate regression curve was drawn by Wolfenbarger (1946). A summarization of the data, in terms of average number of bees per count, in different directional quadrants and at different distances from the apiary is given in table 37.

Table 37

Honeybee abundances at distances and in directional quadrants

Quadrant	*Distance from apiary (miles)*						
	0.75	*1.00*	*1.25*	*1.50*	*1.75*	*2.00*	*Total*
Northeastern	1.7	0	0.9	0.6	0	0.3	3.5
Southeastern	1.7	0.2	0.4	0.3	0.1	0.1	2.8
Southwestern	2.0	0.9	0.3	0.7	0.1	0.4	4.4
Northwestern	8.7	5.2	2.4	2.6	5.4	2.9	27.2

More bees were found in the northwestern direction than in any other. Although bees fly from an apiary in all directions, at times they are more abundant in some directions than in others, according to Eckert (1933). Honeybees have a tendency to return to the same field on successive days for foraging. This behavior will sometimes explain unequal distribution in some directions.

Directional influences on passive disperser organisms are understandably dependent on the agent dispersing them. Rivers, creeks, and ocean currents maintain consistent and dependable unidirectional movements. Air currents, although prevailingly directional in some places, are inconsistent and changeable in most places. Dispersion studies with balloons by Felt (1937) and Gaines and Ewing (1938) and with fungus spores by Stepanov (1935) and Wilson and Baker (1946a) covered comparatively short time periods, a few minutes to a few days during which air currents presumably prevailed in general directions. Directional effects during prevailing air movements and for short periods of time might be expected. If the incidence of some disease as a manifestation or evidence of spore dispersal is determined after a long time period—perhaps after changes in wind, light, humidity, temperature, and other factors—lesser directional effects may be found. Most abundance and condition, light, humidity, temperature and other factors influence the incidence of diseases, however, and knowledge of spore dispersal obtained from studies over a short time period may not be congruent with the incidence of disease resulting from spore dispersal over a long time period. Further discussion is given under wind speeds in chapter 5.

Unidirectional dispersion. Unidirectional flights toward northerly and easterly direction were observed by Felt (1937) and by Gaines and Ewing (1938) from the release of balloons. Additional details supplied by correspondence with Dr. J. C. Gaines permitted classification as to direction of recovery from release points. These follow in table 38.

Table 38
Dispersal of balloons based on directions

Direction	NE	SE	SW	NW
Balloons recovered (No.)	97	25	0	34

A very pronounced northeasterly movement was found. Some balloons moved southeast, some moved northeast to make a wedge-shaped pattern; no balloons moved in a southwesterly direction.

Stakman and Hamilton (1939) noted a definite unidirectional northward springtime movement of stem rust disease spores of wheat, *Triticum* spp., caused by the fungus, *Puccinia graminis tritici* Eriks E. Henn.

They observed a movement of spores from Texas hundreds of miles into North Dakota. A southeasterly movement of spores in the fall was also noted, indicating a return trip. This might be termed migration of a passive disperser organism and as such may be a unique or a practically unknown phenomenon.

These observations were made in the Mississippi River basin, in the great wheat-producing region of the United States lying between the Appalachian and the Rocky Mountain chains. Wheat plants on the millions of acres in this area provide the medium for growth and development of the rust organisms. This medium and the singular inter-mountain position, with its climatic differences and seasonal changes, may provide optimum conditions for the exchange of stem rust spores, depending on the season in which wheat plants are growing.

More knowledge has doubtless been obtained from research work on wheat stem rust than from work on other species of rust. Further research might indicate, however, that spores were dispersed equally as distantly and omnidirectionally from the sources of countless billions of spores in Texas and Oklahoma. Spore trapping results by Stakman and Hamilton (1939) show total spore counts for two square feet taken on June 13 and 14, 1938, with Dallas, Texas, taken as zero distance, in a northern direction, as shown in table 39.

Table 39
Directional dispersion of wheat stem rust spores

City and State	Distance from Dallas (Miles)	Spores trapped two sq. ft. (Number)
Dallas, Texas	0	16,704
Oklahoma City, Okla.	214	179,216
Wichita, Kansas	387	336,000
Beatrice, Nebraska	604	54,336
Brookings, South Dakota	880	12,624
Fargo, North Dakota	1,105	1,344

Spore production by wheat plants in Texas apparently extended from Dallas, Texas, northward to Wichita, Kansas, in such manner that cumulative spore loads of northward moving winds reached a maximum at Wichita, Kansas, and reductions due to spore deposition became more evident from that city to Fargo, North Dakota, 700 miles distant. This is shown in figure 21 a reproduction from Stakman and Hamilton (1939). The deposition of 1344 spores per square foot at Fargo suggests that spore fall continued to some distance northward from Fargo and became zero spore fall at some unknown distant point.

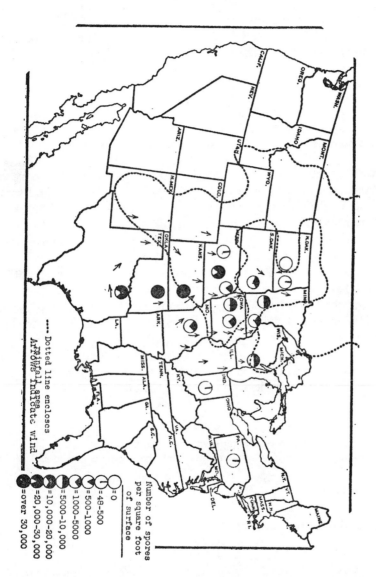

Fig. 21. Unidirectional northward dispersal of wheat stem rust spores from Dallas, Texas. Spores were caught on Vaselined slides exposed June 14, 1938 (fig. 3 from Stakman and Hamilton).

Migratory grasshoppers of two species, *Dissosteira longipennis*
(Thos.) and *Melanoplus mexicanus* Sauss., were released by Willis
(1939) to study the direction and distance of flight. Of 20,440 released,
30 were recovered and are classified as to direction from liberation to
recovery points. The results appear in table 40.

Table 40
Direction of movement of grasshoppers

Direction	N	NE	E	SE	S	SW	W	NW
No. of grasshoppers	3	4	0	0	0	3	0	20

A northwesterly movement is indicated by the largest number recovered
in that direction, few recovered in N, NE, and SW directions and some
in E, S, and W directions. Although a trend is indicated by these data,
the total number, 30, is comparatively small.

Directional influences, dependent on wind direction, were recognized
for "locusts" by Williams (1958). He admitted that "it is difficult now
to believe that the wind direction determines the movements of butterflies and moths."

Grasshopper, *Mecostethes magister*, movements toward the N, E,
NE, and NW exceeded those toward the S, W, SW, and SE, according
to Nakamura *et al.* (1964), following the mark-release-recapture of the
insects. They gave the directions of movement from the dominant to the
recessive directions and ratios of the most to the least with arrows point-
ing toward directions with lesser movement, in table 41.

Table 41
**Mark-release-recapture of a grasshopper
in response to directional movements (data from Nakamura *et al.*)**

	Second and third day	Fourth and fifth day
Cardinal directions	N>E>S>W	W>N>E>S
	4.519	1.881
Inter-cardinal directions	NW>NE=SE>SW>SE=SW>NE	
Ratio	2.123	1.741

An analysis of dispersal and movement in *Phaulacridium vittatum*
(Sjost) was given by Clark (1962). He found that non-random movements of the grasshopper oriented in an elliptical pattern with the longer
axis in north-south directions. Such orientation was suggested as attributable to polarized light or to the sun's position. From the data given
(table 2) for the Majura site and from the March 1958 release of two

groups, displacement distances were given for days' movements, expressed as root mean squares (yds.) in table 42.

Table 42
Orientation of released grasshoppers in north-south directions

Days after release	1	5	11	18	22	25
Root mean squares (yds.)	7.431	3.456	2.696	3.080	3.515	3.678

The principle illustrated by these data is that an organism disperses more rapidly in the initial and more slowly in the terminal stages of the dispersion journey. Such changes in rate may be attributed to aging of the organism, although other and unknown factors may be operative.

Other examples of the principle follow. Comparatively few sets of data illustrate the principle involved and some of these lack perfect alignment. Problems inherently involved may exceed those usually encountered in studies on dispersion.

Unidirectional dispersion is thus indicated for beet, potato, and six-spotted leafhoppers toward the north. (Would related species disperse southward if released south of the equator?) V-shaped patterns are indicated as the form of distribution. Summertime movement of the Mediterranean fruit fly, *Ceratitis capitata,* up the Rhone valley into central Europe may be a similar example of dispersal. Undoubtedly many other examples of unidirectional dispersion occur.

There is a likelihood that atmospheric conditions over the Missouri-Mississippi river valley system are adapted to making favorable conditions for northward dispersal (perhaps also in a southward direction). Wheat stem rust spores move regularly from areas of production northward, as discussed above. Active-disperser organisms may also move in like manner.

Leafhoppers, although they may strive to maintain control of their movements may lose control of their flights and be "unable to help themselves." They may be caught by turbulences and dispersed northward to terminate their journey hundreds of miles from the source. Unidirectional dispersion may be a response to conditions that favor the organisms' lives and abilities to reproduce. Journeys in other directions, one or more of which may be unknown to man, may be hindered or barred. Another explanation, purposive disposal, might be connected with or related to true migration.

Unidirectional dispersal of the beet leafhopper, *Circulifer tenellus* (Bak.), in which the insects moved in a northerly direction from the winter breeding grounds, was first suggested by Carter (1929), later by

Annand (1931), and later described by Romney (1939) and illustrated in figure 22. Surveys, ecological investigations, and general observations determined leafhopper and host conditions that enabled the movements to be traced as far as 430 miles. A perennial mustard, *Lepidium alyssoides* A. Gray, the principal source plant, grows along the Rio Grande River in New Mexico and Texas.

An apparent unidirectional movement by the potato leafhopper, *Empoasca fabae* (Harris), was suggested by Poos (1932), De Long and Caldwell (1935), and more recently by Medler (1960). Early spring-

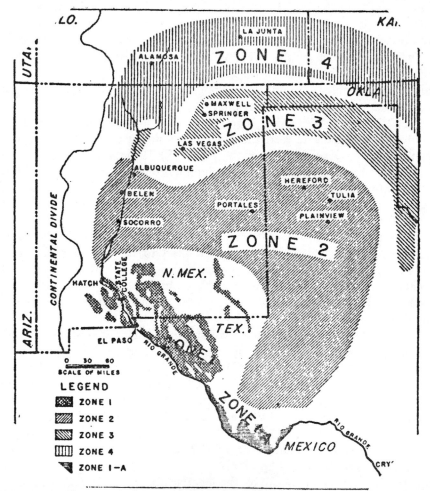

Fig. 22.   Zonation effects of sugar beet leafhopper dispersion from over-winter areas of the insect originating along the Rio Grande River in Texas and southern New Mexico (fig. 3 reproduced from Romney).

time populations of adults in the Gulf Coast states or in the lower Mississippi valley develop and, presumably, begin movements northward. More data are needed, however, to provide evidence of directional movements.

A similar northward movement is achieved by the six-spotted leafhopper, *Macrosteles fascifrons* (Stal), according to Drake *et al.* (1965). Source areas of the insect and its northward dispersal patterns drawn for central North American states are given in figure 23.

Left or right turns. Left-left or right-right turns would result in curvature of the dispersal journey patterns. The distance of movement between turns and the angle of the turns could indicate whether complete circles or wide arcs resulted. The size of the circles, whether wide, covering hundreds of feet, or short, covering perhaps ten feet, would depend also on the angle of turn and the distance of movement between turns. Various observations suggest, however, that various and inconsistent angles are turned and different distances are covered. Heterogeneous movements, large and small circles, and large and small arcs, appear well expressed by the term *meandering.* These movements depend on the species, its dispersal phase or activity, and the factors or conditions involved that are encountered at the time of and during movement.

Unidirectional dispersion may be more frequent and sustained for active than for passive disperser organisms. Data that give comparisons of the two types, however, are not known. A generalized concept is that an individual turns, or is turned, in its dispersion journey. One stimulus or more is usually believed to motivate changes in direction between origin and terminus.

Horizontal dispersion of grain beetle movements were measured according to left or right turns in a thin sandwich of wheat grains between two sheets of glass. Data were given by Surtees (1964) for three species and from laboratory studies, as shown in table 43.

Table 43

Left and right turns of the sexes of grain beetles

Species	Sex				
	Male			Female	
	Left	Right		Left	Right
Sitophilus granarius	32	27		34	27
Oryzaephilus surinamensis	18	23		15	17
Tribolium castaneum	17	35		21	24
Totals	67	85		70	68

Totals for the sexes of three species	Left	Right
	137	153

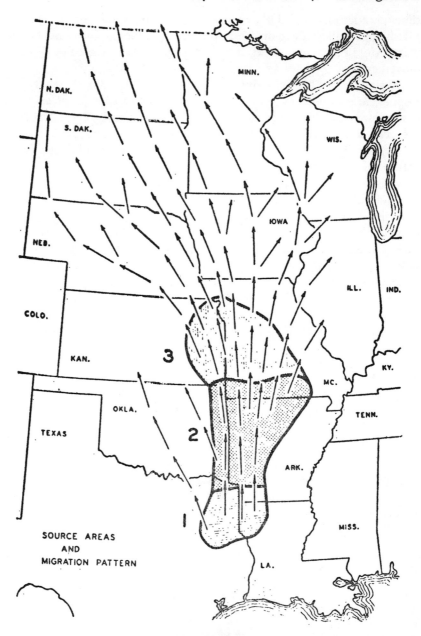

Fig. 23.   Source areas and dispersal pattern of the six-spotted leafhopper from the source areas 1, 2 and 3, which are successive areas of production as the dispersants move northward into the north central states (fig. 9 from Drake *et al.;* reproduced by permission of the Entomology Department, College of Agricultural and Life Sciences, University of Wisconsin).

More right than left turns are indicated by the totals for both sexes for the three species. More right than left turns are indicated for males than for females. Two more left than right turns were found for females. These data are probably statistically insignificant, actually showing random responses to directions.

Directional influences of foraging bumblebees on radish plants were studied by Bateman (1947b). These data are tabulated and indicated in table 44.

Table 44
Directional turns of foraging bumblebees on radish

Subsequent flight	First flight movement	
	Left	Right
Left	55	42
Right	42	60

Directions in the first flights were followed by subsequent ones in 115 movements of 199. These continued directions were non-random and in statistical terms were significant. Differences between the turns, accordnig to the writer, however, were "not very great."

Latitude. Dispersion is apparently independent of latitude. "The dust content of the atmosphere seems to be distributed over the different latitudes very nearly in proportion to the total energy income through radiation from sun and sky," according to Angstrom (1930). Dust content is presumed analagous to organisms in this consideration. Organisms differ with latitude because of locality differences by species and by densities.

Coriolis effect. Coriolis forces would tend to move organisms to the right in the northern hemisphere and to the left in the southern hemisphere, as the force of the earth's rotation moves them toward the poles.

The distribution of species according to biogeographical studies has occurred from dispersion that may be attributed to coriolis effects. Species gradients were observed by Briggs (1967) and Fell (1967) as they described species distributions around the world that proceeded in a westerly direction.

The western Atlantic tropics are a secondary center of the evolutionary radiation of tropical shore fish, *Entomacrodus* sp., according to Briggs (1967). Species of the fish originating in the eastern Atlantic are incapable of successfully invading the western side. There are sixteen or seventeen species of *Entomacrodus* in the western Pacific, six or seven in the Indian Ocean, two in the eastern Atlantic, two in the western Atlantic and one in the eastern Pacific. The origin of the genus is considered to be in the western Pacific, from which species migrated to the New World across the Pacific, thence eastward across the Atlantic to West Africa.

A gradual diminution in the number of marine species of molluscan species of the genus *Voluta* around the world in a westerly direction was reported by Fell (1967). He believed that speciation intensities observed are the results of coriolis effects and that this offers a plausible explanation for certain observed relationships.

Definitive observations on the distance of dispersal of organisms attributable to coriolis effects appear lacking. Many instances of movements northward and some southward could be cited, for example: locust invasions in many countries; Lepidoptera so well described by Williams (1958), leafhoppers, and aphids from the southernmost into the more northern areas of the United States and southern Canada; dispersal of wheat stem rust spores from the southern into the most northern states and the Canadian provinces. Are these movements the results of earth, sun, water, or other factors? Or could these poleward dispersal or migratory movements be in response to coriolis effects?

Restricted directional dispersion. Directional influences by agents of dispersion may be very positive, as in the case of rivers or other water currents. Roads or highways are also positive; through highways may be strategic directions or avenues over which organisms are dispersed. Railroads also remain positive, and main lines especially may provide transport for organisms. Examinations of railway cars permitted Campos *et al.* (1961) to give 0.79 as the average number of mosquitoes per box car that ended dispersal at a terminal, Matamoros, from the interior of Mexico.

Spores, seed, pollen or other kind of organisms that are dispersed in this way are termed hitchhiking organisms. Considerable discussion often is given to hitchhiking organisms—those dispersed inadvertently by airplanes, automobiles, ships, trains, and other vehicles of transportation. Although there is no question that organisms are dispersed by such moving objects, there are questions concerning the frequency, directions, amounts, and distances of such movements. One reference gives definitive data on distances to which hitchhiking occurred.

Following tsetse fly studies in Kenya, Lewis (1950) presented data on transportation of the flies by railway trains. *Glossinia longipennis* Corti, *G. pallidipes* Aust., and *G. austini* Newst. were the species taken, in that approximate order of abundance. Flies entered trains enroute to Nairobi from Mombasa. Flies entered trains along the way, with Mtito Andrei taken as the origin of the hitchhiking flies. Data are tabulated in table 45.

Table 45
Hitchhiking tsetse flies on trains in Kenya

Miles from Mtito Andrei	31	62	103	122	140
Number flies per train					
Goods train	186	103	46	6	6
Passenger train	6.7	3.0	2.0	6.3	1.0

Many more flies were taken from goods trains than passenger trains, more especially nearer the fly sources. It is not known whether the higher speed of passenger trains or their less favorable hiding places may account for the differences between trains. Whether conditions were favorable or unfavorable, the flies rode for distances in excess of one hundred miles.

Changes in the modes of man-made carriers of organisms were discussed by Joyce (1961), who observed that surface water craft were formerly of importance in the dispersal of organisms to Hawaii. Previous to World War II many islands in the Pacific area were reportedly free of mosquitoes. They became infested, however, and aircraft were believed at least partially responsible for carrying mosquitoes to them. Examinations showed (Joyce, 1961) that 3.3 percent of the mosquitoes on planes from foreign airports were alive. Of 2,341 aircraft examined by Joyce (1961) in 1955-59, 56.5% were carrying insects, with 8.19 the average number per aircraft. Of 246 aircraft observed by Laird (1951), 36 percent had insect stowaways of 548 insects, representing 56 families, 10 orders. There were nine spiders.

Localized dispersal of seeds by birds and other animals may be seen in various places and manners. Examples noted in southern Florida are royal palm plants growing beneath avocado trees not far distant and tomato plants surrounding tomato fields. These are seen as mute evidence of dispersal by birds or other animals, similar to the discussion by Krefting and Row (1949).

Solar effects. Students have long discussed the way or ways in which insects orient themselves during migration. Although many possible factors have been considered, perhaps evidence was insufficient to prove that any particular means was employed. An analysis by Baker (1968) on the flight direction of eight species of butterflies showed that five oriented themselves by the sun without compensating for its diurnal movement. The sun, with its effects of heat and light, is a powerful factor affecting the movement of organisms. Except as evidence is obtained to the contrary, other factors may be insignificant in comparison to solar radiation.

Gravitational Factors

4

Gravity affects dispersion of all organisms in one manner or another, whether in movement (from one spot to another) or in remaining quiescent. Those organisms that move to lower elevations may do so in response to gravity. Those organisms that disperse to higher elevations may be resisting gravity in such movement. Studies on the elevation to which organisms disperse or are dispersed are reported by various authors. It is often implied or suggested that organisms in great abundance at high altitudes, i.e., having high vertical movement, disperse to great distances horizontally. There may be a relationship between the elevations to which organisms disperse and the horizontal distances to which they travel, but no relationship appears to have been determined. There is also an implication that organisms that remain near ground level must disperse for only short distances.

Inorganic matter and organic units—spores, seeds, pollen, and insects —must obey the laws of earthly gravitational attraction. Gravitation is recognized, has been measured, is everywhere, is dependable, and yet remains an enigma, according to Dicke (1959). Sufficient energy must arise from within or without organisms to overcome gravity and affect movement, whether they live in air, water or soil. Variations in the size, shape and density of organisms influence movements in any medium. Studies of the rate of fall of spores by Buller (1924), Schmidt (1925), Stepanov (1935), Ukkelberg (1933), McCubbin (1944), and Gregory (1945) give evidence of the complexities involved. Studies of the rates of ascent of small organisms appear lacking, but the complexities involved are probably as prevalent as are those involved in descent. Turbulence and convection currents can transport organisms and provide energy transfers that overcome gravity.

Gravitational forces are of value in affecting the departure and return of organisms from and to the earth's surfaces. A process is thus provided for maintaining the earthly positions of organisms. Although dispersive forces move organisms, opposing gravitational forces tend to retain them. Energy for sustained vertical movement must eventually cease for each individual, since it cannot live indefinitely in the air without receiving nutrition from the earth or from an earth-bound object. Gravity retains some individuals and prevents their departure from areas favorable to them, thus affecting the economy of a species.

Atmospheric diffusion at lower altitudes was discussed by Sutton (1947) in connection with chemical warfare. Smoke was described as dispersing (from a point source) in the form of a cone whose dimensions are affected by air turbulences. Surface roughness from vegetation and other physical features causes distortions in cloud patterns within the cone. Theories of movement by the wind lead to the correct functional form only when extreme values are admitted for the turbulence factor, according to Sutton (1947).

TURBULENCE

It is a common observation that movements of suspended particles, such as smoke, although continuous from the origin, proceed with a lack of uniformity and in disorderly patterns. Initial particle movements may (and frequently do) appear equal at the outset of the dispersion process, as discussed by Corrsin (1961), but later indicate turbulent flows. Smoke streaming from an origin in a very calm period, for example, may ascend in a *laminar* flow pattern for some distance (fig. 24). Then, owing to unequal pressures or flow of air movements, turbulences begin. The production of eddies render the flow patterns unequal in speed and direction of movement. Sampling of air in the paths of the turbulent smoke columns at any given time would reveal wide variations in smoke particle densities owing to eddies. The more disturbances contacting the laminar flows, the shorter is the laminar phase of dispersion (Corrsin, 1961). As the distance from the smoke origin increases, the particle densities become sparser, to the point that they are no longer observed.

Organisms, especially the larger species, may be observed occasionally in such densities that turbulence and dilution may be recognized in the dispersion process. Usually organisms are so few, however, that the relationships are not observed. Laminar patterns of dispersal of organisms may exist initially for short distances from the origin. As distances increase from the origin, unequal microclimatic and other factors along the dispersal routes retard and deflect some organisms, while others proceed undisturbed at the original speeds and directions. Among active disperser organisms there are undoubtedly repellent and attractive stimuli that serve to induce turbulence.

Although laminar and turbulent movements may not be observed, they may be expected and considered. Sampling organisms in the laminar phase may be less variable than in the more distant turbulent phase. Determination of the regression or other statistical measure may be more

Fig. 24.   Laminar stream of smoke from a cigarette extends about one-half the distance from cigarette to the top of the picture before it becomes turbulent and terminates in eddies (from an article by Corrsin in *Johns Hopkins Magazine*. Photograph by Werner Wolff—Black Star).

positive with less sampling in the laminar than in the turbulent phase. Also, it is possible that the rate of movement may differ somewhat, depending on the conditions of laminar or turbulent dispersal. Turbulence may explain some variations and give some understanding of certain observations in some regression studies. Turbulence will almost certainly explain many variations in the more distant dispersal ranges of an organism.

Further understanding of dispersion may be obtained from studies of smoke, gas, and dust movement and soil erosion. A lack of basic data was reported by McKenzie (1958) with reference to air pollution in his expression of concern for public health. Special reference is made to smoke, gaseous vapors, chemicals, and other particulate emissions. A note of warning was given by Hagen-Smit (1958) on conservation of the air containing pollutants. Such contaminants are increasing with the years, on a real basis owing to increasing populations and on a per capita basis owing to newer modes of transportation and higher standards of living. Studies of changes in concentrations, wind speeds and turbulences, and chemical reactions were recommended in order to determine trajectories and isopollution lines.

Very rapid and general or widespread multiplication of facultative disease-producing organisms among a population of previously healthy individuals may be termed *epidemics*. Unless these organisms disperse or are dispersed there is no epidemic. Wells (1955) discussed contagious organisms, hygiene, and sanitation, indicating the significance of dispersal as related to epidemics. Airborne infections were also discussed by Rosebury (1947) from studies on contaminants. Epidemics tend to be confined or to affect organisms in rather restricted areas. Organisms of epidemics disperse, therefore, over comparatively short distance ranges. Ecology of invasions by animal and plant species over longer distances and to other continents was discussed by Elton (1958).

Inorganic dispersers. Much work has been done on soil erosion, which is a dispersion process. Comprehension of principles affecting movement of soil particles may assist in better understanding of organismal dispersion. Factors influencing soil erodibility by wind are: (1) size, shape, and density of erodible fractions; (2) size and total of nonerodible fractions; (3) soil moisture and rainfall effects; and (4) mechanical stability and abradibility of soil structural units, according to Chepil (1958). Certain of these, especially size, shape and weight, are recognized as important factors. Soil particles, in obedience to gravitational forces, are affected by wind forces striking the earth's surfaces. The forces are reduced by soil particles through "drag" effects. Smaller particles, such as smoke or haze, remain suspended longer and are less susceptible to displacement by air molecules than are larger particles. Gravity is counterbalanced, according to Bagnold (1942), by the dispersive tendency of movement energy that the smaller particles receive from air molecules.

Bagnold (1942) further concluded that: half-speed reductions of smaller particles by air drag require less time than that of larger particles; sand grains travel downwind and approach, but never quite attain, the speed of the wind carrying them; sand grains fall more slowly than they arise because of air drag; deposition of sand grains in their final resting places is induced by sedimentation, accretion, or encroachment. In *sedimentation* and grains fall through moving air with insufficient velocity to be carried forward. In deposition by *accretion* sand grain movements are reduced if (a) the surface wind becomes feeble or (b) the surface changes in texture. In deposition by *encroachment* sand grains come to rest by striking an obstruction. Deposition of most organisms probably occurs through sedimentation although accretion doubtless accounts for fallout.

Vertical profiles of spores and pollen were illustrated by Hirst *et al.* (1967), suggestive that much remains to be learned about vertical dispersion. Extremes of organismal movement are apparently equal to extremes of the various turbulences of air movements. Air "boils," manifested by cumulus clouds, are also suggestive of vertical movements present in the air above the earth. Although there must be energy to move the organisms and a lag doubtless occurs, opportunities must be very great for high vertical and horizontal movement for the aerial dispersal of many organisms. One might ask: "How does it come that no more worldwide dispersion occurs than appears evident?"

VERTICAL DISPERSION

A relative abundance of material and organisms, inorganic and organic, has been reported at various altitudes by many workers.

In the vicinity of Washington, D. C., Kimball and Hand (1924) reported the presence of many dust particles, at different elevations, as shown in table 46.

Table 46
Dust particles at various elevations
(data from Kimball and Hand, 1924)

Month counts were made	*Altitude (feet)*					
	0	1,200	4,000	6,000	8,000	10,000
August	232	313	308	250	81	60
October-November	357	282	157	94	63	43

Although decreases in dust particles were found with increases in elevations, a not unexpected occurrence, the significant feature is the comparatively large numbers of dust particles at ten thousand feet. Termination of dust particle dispersion, moreover, is at some elevation above ten thousand feet. Many dust particles were present in each of the sampling periods, suggesting that many dust particles are constantly present above Washington, D. C. and, most likely, other localities.

ORGANISMAL DISPERSERS

An abundance of bacteria and mold was reported by Proctor (1934) from 1,500 to 15,000 feet elevations. The expected numbers of these organisms were obtained by regression formulae by Wolfenbarger (1946) (table 47).

Table 47
Bacteria and mold spores at elevations
(data from Proctor, 1934)

Organism	Altitude (feet)					
	1,500	*3,000*	*6,000*	*9,000*	*12,000*	*15,000*
Bacteria	113	80	48	29	15	5
Mold	106	70	34	12	0	—

More bacteria than mold organisms were displaced and moved to higher elevations. Mold organisms were present to more than 9,000 feet, bacteria to more than 15,000 feet. In sampling air for mold, MacQuiddy (1935) reported innumerable counts at 3,000 and 4,000 feet, respectively. Bacterial colonies numbered 423, 36, 51, 2, and 7 at 3,000, 4,000, 5,000, 6,000, and 7,000 feet, respectively. Pollen grains per square centimeter numbered 167, 38, 17, 11, and 9 at 3,000, 4,000, 5,000 and 7,000 feet, respectively. In his table 2, MacQuiddy (1935) recorded bacteria and pollen upwards to 7,000 feet, the numbers of which are given as expected numbers from Wolfenbarger (1946) in table 48.

Table 48
Bacteria and pollen at various elevations
(data from MacQuiddy, 1935)

Organism	Altitude (feet)					
	1,000	*2,000*	*3,000*	*5,000*	*6,000*	*7,000*
Bacteria	193	167	152	132	126	120
Pollen	121	13	9	3	1	0

More bacteria than pollen were displaced, and proportionately so, to higher elevations. The reduction of bacteria was from 193 at 1,000 feet to 120 at 7,000 feet, a comparatively small difference.

Non-spore-forming bacteria were reported by Mischustin (1926) to disappear between 1,000 and 2,000 feet. He reported the number of bacteria per liter of air as 2.2, 4.2, 1.7, and 0.6 at 200, 1,000, 1,500, and 2,000 meters respectively. The data show decreases to 2,000 meters and suggest that bacteria are present to some distance in excess of 2,000 meters. Petri dishes were used in making the collections, and data were given on bacteria per petri dish and converted to bacteria per liter of air. The resulting standardization enables others to continue this research.

Bacteria and fungi collected by Wolf (1943) were reported by graphic illustrations, from which the number of organisms per cubic foot are given in table 49.

Table 49

**Bacteria and pollen at various elevations, on each of
two collection dates (data from Wolf, 1943)**

Collection date	Altitude (feet)					
	1,000	2,000	3,000	4,000	5,000	6,000
Dec. 6, 1941	0.68	0.23	0.18	0.03	0.02	0.02
Jan. 25, 1942	0.34	0.20	0.13	0.10	0.08	0.06

Comparatively more organisms were collected at lower altitudes on
December 6 than on January 25, which suggests that the numbers of
airborne organisms vary from one day to another.

Fungi. Many data show the elevations to which fungus spores have
been collected. The significance of vertical dispersion lies in the frequency with which spores return to earth in viable condition. Many
spores are airborne; many are dispersed by agencies other than air.
Because of such widespread vertical dispersion, intercontinental dispersion may occur. Global exchanges of certain species of organisms—
for example, rust spores, mites, and aphids—may occur infrequently.
Important questions, however, may arise from consideration of such
widespread dispersions. One might ask: "Should not certain species have
worldwide distribution?" "How can certain host plants be grown without infestations?" and "What is the significance of quarantine measures
against the introduction of organisms from foreign lands?"

Onion mildew, *Peronospora distructor* (Berk.) Gaspary, spores decrease in abundance with an elevation increase, according to Newhall
(1938). This increase extends from 50 to 1,200 feet elevation, with an
apparent exception at 1,500 feet, as given in table 50.

Table 50

Spore spacings at various elevations

Elevation (feet)	50	100	200	700	1,200	1,500
Cubic feet of air/spore	32	46	40	467	800	171

A straight-line regression curve was indicated by the data, according to
Wolfenbarger (1946), through omission of the 1,500 foot elevation record of 171 cubic feet of air per spore.

Alternaria spores were taken in airplane flights and by gravity at
ground level by Vinje and Vinje (1955). The mean number of spores
from six collection days' flights, each of two minutes exposure, are given
in table 51.

Table 51
Spores collected per two-minute period at various elevations
(data from Vinje and Vinje, 1955)

Alt. (ft.)	0	500	1,000	1,500	2,000	2,500	3,000	3,500	4,000	4,500	5,000
Spores (no.)	20.4	7.8	5.6	3.8	7.2	6.2	8.8	5.0	5.3	5.2	4.2

Most spores were taken at ground level, the second rank number at 3,000, and the third rank number at 500 feet. A very slight gradient is indicated for the range 500 to 5,000 feet elevation, although zero number of spores was at some elevation in excess of 5,000 feet.

In studying the forecasting of rice blast disease spores, *Piricularia oryzae,* Kono (1965) showed relationships between the number of spores caught and the height of the spore traps, operated twenty-five days, at distances above ground (table 52).

Table 52
Number of rice blast spores caught by traps
at distances above ground (data from Kono 1965)

Height of trap above ground (m.)	0.9	3.0	6.0	9.0	12.0	18.0	21.0	24.0
Number of spores	4,419	706	516	363	321	202	154	101

Distribution of spores was based on the "power law," according to Ono (1965), meaning that the nearer to earth the traps were, the more spores were caught. Also, in regard to the difference in numbers of spores at different heights, "the stronger the wind the less the difference."

A straight-line regression curve was indicated, according to Wolfenbarger (1959), for the data on vertical abundance of yellow rust, *Puccinia glumarum* (Schmidt) Erikess. & Henn., reported by Hubert (1932), in table 53.

Table 53
Yellow rust spore abundances

Elevation (meters)	34	400	600	800
Spores per sq. cm. per min.	1418	683	336	62

Straight-line relationships may be attributed to the comparatively low elevations in the above observations or they may have been in the optimum zone of dispersion. Heights to 1,200 feet (since the 1,500-foot elevation appears out of line) given by Newhall (1938) and to 2,600 feet by Hubert (1932) may have been insufficient, possibly in combination with variants in the number of spores taken, to give the curvature common to most vertical dispersion observations. On the other hand, portions of the spores may have been killed by high altitude the time spent in dispersion at high attitude, alone with other factors providing the relationships indicated.

Apparently more efforts have been expended in studies on the dispersion, vertical and horizontal, of wheat stem rust, *Puccinia graminis* Pers., than on any other species of fungus. Such efforts doubtless have been motivated by (1) the great losses from wheat stem rust, and (2) the striking distance to which stem rust spores are borne by air currents, making dispersion studies on them an intriguing challenge. Data from Craigie (1945) and Peturson (1931) show comparative abundances of wheat stem rust spores in table 54.

Table 54

Wheat stem rust spores at various elevations
(data from Craigie, 1945, and Peturson, 1931)

Elevation (feet)	1,000	5,000	10,000	14,000
Number spores (Craigie)	48,200	7,730	144	40
Number spores (Peturson)	10,180	1,180	28	11

Data reported by Stakman *et al.* (1923) show much less effect by elevation than is indicated by the above data (table 55).

Table 55

Spore abundances at various elevations
(data from Stakman *et al.*, 1923)

Elevation (feet)	1,000	2,000	5,000	7,000	10,000	12,000
No. spores	20	12	8	3	4	1

Wheat stem rust spores are shown to disperse to high altitudes, over two miles. The figures suggest that spores must have been dispersed to elevations in excess of the maximum elevations given. In view of these observations one may ask: "May not spores commonly disperse to four, six or ten miles?" and "If spores were present at such altitudes, might not some of them drift horizontally for hundreds of miles, transcontinentally, or even encircle the globe?" Intercontinental distribution of plant pathogens has been accomplished more by man, according to Stakman and Christensen (1946), than by wind, which suggests that high vertical dispersion is an hypothesis more than a reality in accomplishing wide horizontal spore movement.

Considerable data were given by MacLachlan (1935) on the abundance of basidiospores of *Gymnosporangium juniperi-virginiana* Sch. and of *G. globosum* Farl. at altitudes above the sources. Spore collections, made at comparatively low altitudes, are indicated in table 56.

Table 56
Abundances of G. *juniperi-virginiana* basidiospores
(data from MacLachlan, 1935)

Elevation, feet	100	500	1,000	1,500
Number of spores, average	19.5	10.0	2.0	0.0

Pollen. Pollen grains must be dispersed in order to pollinate ova and to induce allergies. Gravitational forces must be surmounted to move pollen to accomplish these objectives. Ragweed, *Ambrosia* spp., pollen air-index averages ranged from 44.1 at Toronto, Canada, to 0.1 at 10 miles southwest of Fort Williams, Canada, indicating wide variations between locations, according to Bassett (1959). It is expected that most pollen would be found nearest the sources.

Mean diameters of *Betula* sp. pollen grains were less at higher than at lower altitudes, according to Rempe (1937). More pollen grains were found by Rempe (1937) during day than night (table 57).

Table 57
Betula pollen abundances in day and night flights
(data from Rempe, 1937)

Elevation, meters	10-40	200	500	1,000	1,500
Day flights	904	849	852	581	267
Night flights	577	560	283	85	45

Pollen count decreases were found with elevation decrease from day and night flights. Regression studies, based on straight-line and curvilinear relationships, indicate similar rates of decrease with elevation decrease from day and night counts.

Airplane flights were used by Meier and Artschwager (1938) to trap sugar beet, *Beta vulgaris,* pollen grains on agar plates. These data are given in table 58.

Table 58
Sugar beet pollen at elevations
(data from Meier and Artschwager, 1938)

Elevation (feet)	1,000	2,000	3,000	4,000	5,000
Number pollen grains	56	25	14	9	28

A decrease in pollen grains with elevation increase is seen except at 5,000 feet. The 5,000-foot elevation was in the "dust zone," according to the authors. An explanation for the high count of pollen grains at 5,000 feet may be (1) a variant of sampling or (2) a collection from an air stream from a different source and direction from that of lower elevations. (Cross currents of air are known at different altitudes, since cloud formations are seen occasionally in which one stratum moves in one direction while the other stratum moves in another direction.) A high count from insect collections given by Hardy and Milne (1938a) is suggestive of stratification. Between 750- and 1,000-foot elevations 4,400 insects were taken; the count was 16,700 insects for the 1,000- to 2,000-foot elevation. Two zones of the atmosphere, based on insect populations, were recognized by Berland (1935): the "terrestrial" below about 200 feet, and the "plankton" above 200 feet. Upward to nearly 200 feet the atmosphere served as a medium for the larger and more active disperser insects. Above about 200 feet the atmosphere contained the weaker, smaller and more passive disperser insects. Two zones, the "terrestrial" below 150 feet and the "aerial" above 150 feet, were recognized by Freeman (1945).

Another explanation for dispersal stratification is inversion layering, which is attributed to different temperatures. Also, the time at which samples were taken, in relation to other factors, may have influenced results.

Microbiotic samples obtained at altitudes by Vinje and Vinje (1955) were studied to note the relationships of sample organisms with weather conditions. Pollen grains were the most conspicuous organisms obtained. Species of Gramineae were most numerous in samples taken on six collection days in two-minute collection time units. The mean of the collections are given in table 59.

Table 59

Gramineae pollen at elevations
(data from Vinje and Vinje, 1955)

Hgt. (ft.)	0	500	1,000	1,500	2,000	2,500	3,000	3,500	4,000	4,500	5,000
Pollen (no.)	36.5	21.2	14.8	13.8	22.8	18.0	14.6	15.4	9.4	10.0	12.0

Although most pollen was found at the lower altitudes, the reduction at 5,000 feet was rather small. This suggests that Gramineae pollen may disperse to very high altitudes.

Seeds. Seeds were collected in the upper air by Glick (1939). Eleven genera in three families were represented in the collections. The total seeds from all species are given in table 60.

Table 60
Abundances of seeds at elevations
(data from Glick, 1939)

Elevation (feet)	200	500	1,000	2,000	3,000	5,000
Seeds (number)	13	2	16	9	9	4

Although these seeds were reportedly collected during periods when the upper air was rough to slightly rough and convection currents were rather strong, the data suggest rather distant dispersion by some seeds.

Nematodes. Vertical dispersion of the potato root eelworm, *Heterodera rostochinensis* Woll., was observed by Wit (1952). Viable cysts were found at elevations of four feet, and the regression curve (Wolfenbarger, 1959) indicates that a few cysts must have been lifted to elevations in excess of four feet. Observed numbers are given in table 61.

Table 61
Eelworm cysts at elevations
(data from Wit, 1952)

Elevations (inches)	6-22	23-38	39-52
Cysts (number)	2,077	435	103

Dispersion vertically exceeds fifty-five inches and terminates at some unknown height.

Airplane flights by Vinje and Vinje (1955) and slides at ground level collecting organisms by gravity were used for sampling dispersing organisms. Samples were taken on six days, two minutes each day, with the mean number of nematode eggs taken at elevations shown in table 62.

Table 62
Nematod eggs at elevations to 5,000 feet
(data from Vinje and Vinje, 1955)

Alt. (ft.)	0	500	1,000	1,500	2,000	2,500	3,000	3,500	4,000	4,500	5,000
Eggs (no.)	14.0	48.0	10.7	10.3	8.3	14.0	7.7	9.7	4.5	8.0	6.0

Most nematode eggs were found at 500 feet and tended to decrease to 5,000 feet. That the number of nematode eggs is one-sixth the number taken at 500 feet indicates high altitudinal dispersal and is surprising.

Mites. Upward movement of the citrus red mite, *Panonychus citri* (McGregor), was reported by Tashiro (1966), with the recovery data given in table 63.

Table 63

Upward, horizontal, and downward movements of the citrus red mite
(data from Tashiro, 1966)

	Numbers of mites recovered in directions at distances		
Feet from release	*Upward*	*Horizontal*	*Downward*
<1	47	15	12
1	45	17	16
2	46	16	2
3	17	5	4
4	1	1	1
5	3	0	2
6	1	0	0
7	0	1	0

Upward dispersal in most instances was about three times as great as downward dispersal. Horizontal movement was less than upward but more than downward movement. If < 1 and 1 are combined, a gradient is obtained that is more characteristically observed in dispersal regressions.

Insects. Investigations of "the insect fauna of the upper air is of both scientific interest and economic importance," according to Glick (1939). Many "small weak-flying insects of high buoyancy" were found by Freeman (1945) drifting involuntarily with the wind, which Freeman contrasted "with the migration of moths, butterflies and locusts, which are sufficiently powerful to maintain their flight if necessary against the wind." Discussions of vertical dispersion apparently presume that flying insects surmount gravity. The degree to which flight displaces gravity is of interest, although air currents and other agents of dispersion may promote or retard flight activities. The heights at which insects are collected becomes the effects of flight, air current movements, and other dispersal agents. Many records are given on the abundance of insects collected at different elevations, however, and some of these are repeated below.

Collections of insects at different elevations have been taken by means of nets or screens that function as sieves. Large volumes of air are filtered through the nets or screens to obtain whatever insects are present. Many species were taken in the collections, which, after classification, were pooled as "total" insects. Discussion will be given first to taxonomic groups, combining several to many species, then to single species.

Based on the number of insects caught with a three-foot diameter net per ten hours of flying time, Hardy and Milne (1938) gave the data in table 64.

Table 64
Abundances of insects collected at elevations
(data from Hardy and Milne, 1938)

Elev. (ft.)	10-150	150-300	300-500	500-750	750-1,000	1,000-2,000
Total no. insects	23,100	8,400	7,100	4,900	4,400	16,700

A total of 24,700 insects was collected by Glick (1939) in 53,633 minutes of flying time at elevations ranging from 20 to 1,600 feet. The average number of insects, all orders, per ten minutes of flying time at different elevations is given in table 65.

Table 65
Average numbers of insects collected at elevations
(data from Glick, 1939)

Elevation (feet)	25	200	1,000	3,000	5,000	10,000	12,000	15,000
Insects (avg. no.)	25.9	13.0	4.7	1.4	0.6	1.3	0.7	0.5

Owing to the need for making comparisons between collections at different elevations, some unit is required for an understanding. The unit used by Hardy and Milne (1938) was "ten hours' time" and by Glick (1939) it was "ten minutes' flying time." The unit given by Freeman (1945) was "number of insects per 1,000,000 cubic feet" of air. Two further differences were observed: (1) the use of moving nets (on kites or airplanes) by the first two references, and (2) the use of fixed nets (on towers) by Freeman (1945). A summary of the total insects collected by Freeman (1945) is given in table 66.

Table 66
Abundances of insects at elevations
(data from Freeman, 1945)

Elevation (feet)	10	177	277
Insects (per 1,000,000 cu. ft.)	240	45	26

More insects were taken at night than by day, on the average, and there were fewer numbers of flightless forms and fewer recognizable orders. A graphic figure from Glick (1939) illustrates the day and night time differences and altitudinal abundances to 5,000 feet (fig. 25).

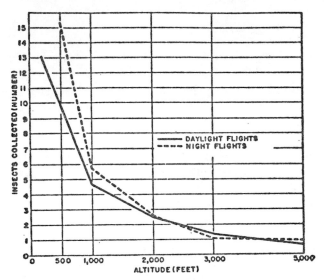

Fig. 25.   Average numbers of insects collected per 10 minutes' airplane flight time, at altitudes of from 200 to 5,000 feet by day and 500 to 5,000 feet by night (reproduction of fig. 4 from Glick).

Vertical distribution of insects is determined to a considerable extent through the productivity of host plants or animals beneath and surrounding the areas where samples are taken. Members of the order Diptera were more abundant than those of any order in the sample collections taken at different altitudes by Berland (1935), Coad and McGehee (1917), Collins and Baker (1934), Freeman (1945), and Glick (1939). Members of the order of Homoptera were proportionately more abundant in the collections by Hardy and Milne (1938, 1938a) however, than were members of the other orders. Such distribution may be attributed to buoyancy of the individuals rather than to abundance of insects on a unit basis at ground level. Highly buoyant insects are disproportionately more abundant at higher elevations than are those of low buoyancy. Such differences led Freeman (1945) to classify and to list certain species as "aerial" and others as "terrestrial." Classification was determined by the frequency with which species occurred at 10,177, and 277 feet. Terrestrial species were those most frequently in the lowest nets and either not at all or only in very small numbers in the upper nets. Aerial species were those that occurred most frequently in the top and middle nets.

Individuals of the lowly insect group Copegnaths were found increasing with elevations to heights of 2618 cm. Adhesive material was used by Profft (1939) at different heights above ground in fields of rape, *Brassica* sp., for taking insects. Observed values are given in table 67.

Table 67
Numbers of Copegnaths at elevations
(data from Profft, 1939)

Height above ground (cm.)	1.7	6.4	11.5	16.6	21.7	26.8
Copegnaths caught (no.)	70	138	295	345	420	440

Although increasing numbers of Copegnaths were caught with increasing altitudes to 26.8 cm., there must be some elevation at which the catch would reach a maximum and begin to decrease. It is suggested that the 26.8 cm. height was near the elevation at which most of the insects would be taken and that the Copegnaths were flying at these heights in order to rise above obstructions due to plant growth or to objects on the earth's surface.

Thysanoptera were taken on adhesive by Profft (1939) to elevations of 50.4 cm., as given in table 68.

Table 68
Thysanoptera taken at elevations
(data from Profft, 1939)

Height above ground (cm.)	5.4	12.9	20.4	27.9	35.4	42.9	50.4
Thysanoptera caught (no.)	661	1100	1116	1177	1102	1205	1183

Increased numbers of Thysanoptera were found with elevation increases to 42.9 cm., and the data suggest a maximum catch may have been reached near 42.9 cm.

Catches of aphids on adhesive in a rape field were found by Profft (1939) to have increased to a distance of 26.8 cm., as given in table 69.

Table 69
Aphid catches at elevations above ground
(data from Profft, 1939)

Height above ground (cm.)	1.7	6.4	11.5	16.6	21.7	26.8
Aphids caught (no.)	2	5	12	10	13	21

More aphids were moving at 26.8 cm. than at lower heights. Aphids, as is found in data on the vertical abundances of other organisms, apparently rise above plant heights to avoid striking earth-bound objects.

Collections of aphids by two nets at different altitudes by Johnson and Penman (1951) showed that most aphids were taken at heights of 50 feet, over a range of 50-2,000 feet. A summarization of the data as aphid density of 10^6 cu. ft. is given in table 70.

Table 70
Mean densities of aphids at elevations above ground
(data from Johnson and Penman, 1951)

Height	50	250	500	1,000	1,500	2,000
Mean density						
Uncorrected	161.9	99.7	26.0	9.1	6.6	4.6
Corrected[*]	1584.0	851.9	67.7	12.3	8.7	4.9

[*]Correction made for the amounts of air passing through the screens.

A series of samples of aphid populations was taken by Johnson (1951) at altitudes of 50 to 2,000 feet. Total aphids per two-hour period per 10^6 cu. ft. of air from each of two tables are given in table 71.

Table 71
Aphid abundances at elevations above ground
(data from Johnson, 1951)

Height (feet)	50	250	500	1,000	1,500	2,000
Table 1	1033	609	417	246	194	146
Table 2	283	228	87	30	6	9

Decreases in aphid abundance with elevation increase were common in the data from both tables. These data were not corrected for wind speeds, although differences were observed in which calmness could account for low catches. Convection currents outlined by Wellington (1945a), which reach their maxima in the early afternoon, may disperse diurnal species to maximum heights in the early afternoon and account for greater densities at certain periods.

The vertical abundance of aphids was reported by Coon and Pepper (1968) as results of dispersion in the late fall period, September 22 to November 13, 1968, tabulated in table 72.

Table 72
Vertical abundance of aphids, totals of aphids, totals of all species
(data from Coon and Pepper, 1968)

Height (feet)	6	10	14	18	23	26
Aphids (number)	463	205	199	138	195	93

Nearly one-fifth the number of aphids were taken at heights of twenty-six as were taken at six feet. Zero collection would be at a height much in excess of twenty-six feet.

Ewert and Chiang (1966) used sticky surfaces on the sides of drums, 30 cm. long and 15 cm. in diameter, to trap coccinellids in corn fields at heights listed in table 73.

Table 73
Number of entrapped coccinellids at heights above ground
(data from Ewert and Chiang, 1966)

Species	Heights above ground (feet)			
	0.9	1.8	2.7	3.6
Hippodamia convergens Guerin	36	2	2	1
H. 13-punctata L.	19	0	1	1
Coleomegilla maculata de Geer	11	0	0	0

The authors suggested that the less nomadic species is phytophagous.

Coccinellids were numerous nearest to the ground level, according to Profft (1939). This is shown in table 74-75.

Table 74-75
Coccinellids at heights above ground
(data from Profft, 1939)

Height (cm.)	1.7	6.4	11.5	16.6	21.7	26.8
Coccinellids (no.)	40	20	14	5	5	1

In view of the carnivorous feeding of coccinellids, adherence to ground level habits may be expected.

Considerable data have been taken on the abundance of the sugar beet leafhopper, *Circulifer tenellus* (Baker), at different heights. These heights were comparatively low in all instances. Relative numbers of leafhoppers were shown by Lawson *et al.* (1951) over a range of elevations from 1 to 127 feet, with results as given in table 76.

Table 76
Sugar beet leafhoppers at elevations
(data from Lawson *et al.*, 1951)

Height (feet)	1.0	2.5	15.0	32.0	50.0	100.0	127.0
Leafhoppers (number)	55.9	28.4	14.7	13.2	12.9	12.2	12.1

Gravid leafhoppers are less active and fly at lower elevations than females with undeveloped eggs, according to Lawson *et al.* (1951). A higher percentage of gravid females was taken at 15.0 than at 2.5 or 32.5 feet heights, as shown in table 77.

Table 77
Gravid sugar beet leafhoppers at elevations
(data from Lawson *et al.*, 1951)

Height (feet)	2.5	15.0	32.5
Gravid females (percentages)	7.4	8.1	4.6

It is suggested that 2.5 feet is near ground level and obstructions, and the insects fly higher to avoid objects; 32 feet is higher than the obstructions; while 15.0 feet is high enough to avoid obstructions, yet low enough to avoid buffeting by winds at high elevations.

Parasitized sugar beet leafhoppers were found by Lawson *et al.* (1951) to fly at lower altitudes than nonparasitized individuals. This is shown in table 78.

Table 78
Parasitizations of the sugar beet leafhopper at elevations
(data from Lawson, *et al.*, 1951)

Height (feet)	2.5	15.0	32.0
Parasitized leafhoppers (percentage)	3.5	1.9	0.7

Gravity is thus seen to be more effective in retaining gravid females and parasitized individuals closer to the ground.

Dispersal of the potato leafhopper, *Empoasca fabae* (Harris), from the Gulf Coast states to the north central United States each spring has been discussed and suspected for decades. A more recent concern over such long-distance movement by Medler (1960) and Drake *et al.* (1965) brought further studies. The data from Glick (1939) at elevations are given in table 79.

Table 79
Potato leafhopper abundances at elevations
(data from Glick, 1939)

Altitude (feet)	200	500	1,000	2,000	4,000	5,000
Glick (1939)	6	3	3	3	1	1
(1960)	7	2	8	3	1	
(1961)	21	4	9	10		

Most leafhoppers were taken at 200 feet and tended to decrease at higher levels, but one specimen was taken in excess of 5,000 feet. No known data give rates of the horizontal dispersion of this leafhopper. Since the sugar beet leafhopper is recognized as dispersing to 450 miles, the potato leafhopper may also move distantly.

Populations of the geminate leafhopper, *Callodonus geminatus* (Van Duzee), were found by Nielson (1968) to decrease above six feet elevations above ground. Values, estimated from the author's semi-logarithmic figure 12, are given in table 80.

Table 80
Relative values of geminate leafhoppers' decrease in height above ground
(data from Nielson, 1966)

Height (feet)	6.0	8.6	9.9	11.0	12.3	13.6	14.9
Log values	2.78	2.66	2.60	2.54	2.47	2.40	2.34
Number values	0.189	0.185	0.182	0.180	0.177	0.174	0.171

Decreased numbers were found over the distance range tested. Zero leafhoppers would be found in dispersing to some considerable height in excess of fifteen feet.

Invasion of an area new to a species, or an epidemic, is evidence that individuals have surmounted gravity to make the necessary movement. Some relative vertical elevation or depth must be utilized in the process of invasion. An invasion of the southeastern United States by a western leafhopper, *Homalodisca insolita*, evidently occurred at low elevations. Adhesive coated boards collected more leafhoppers at low elevations, according to Pollard, Turner and Kaloostian (1959), than at higher elevations. This is shown in table 81.

Table 81
Leafhoppers collected at elevations
(data from Pollard, Turner and Kaloostian, 1959)

Elevation (feet)	2	6	10
Number of leafhoppers	1805	68	24

Whether the invasion was by short or by long hops was unknown. Although there is a possibility of dispersal of the species by some means other than flight and of the less frequent movement at higher elevations (10,000 feet), it appears likely that the invasion was accomplished by comparatively short flights.

Increasing numbers of cotton fleahoppers, *Psallus seriatus* (Reut.), were caught by Gaines and Ewing (1938) with an increase of height to 23.5 feet. This is shown in table 82.

Table 82
Abundances of the cotton fleahopper at elevations
(data from Gaines and Ewing, 1938)

Height (feet)	5.5	11.5	17.5	23.5
Fleahoppers (number)	75	79	124	311

Three of the eight specimens of the cotton fleahopper taken by Glick (1939) were at 20 feet, two were at 200 feet, two, at 1,000 feet, and one at 2,000 feet. Three collections, combined with data from Gaines and Ewing (1938), suggest that the peak of fleahopper dispersion occurs above 25 and below 2,000 feet. Insects, obviously, cannot continue to increase in abundance with elevation increase. An interesting consideration is the elevation at which the abundance reaches a peak and begins to decline.

A regression curve by Wolfenbarger (1946), based on studies by Dominick (1943), shows that about one-fifth the number of tobacco flea beetle, *Epitrix hirtipennis* (Melsh.), adults were taken at twenty-three as at twelve feet (table 83).

Table 83
Tobacco flea beetles at elevations
(data from Dominick, 1943)

Height (feet)	12	14	16	19	21	23
Beetles (number)	99	80	38	54	43	11

Lower elevations:

Height (feet: mid-class points)	1.5	3.0	5.0
Beetles (number)	30	51	25

These data show that beetles dispersed at elevations of from one to twenty-three feet and suggest that most beetles moved at elevations of three feet, after which decreases begin.

European corn borer moths, *Ostrinia nubilalis* (Hbn.), converged to light traps nearer to the ground in greater numbers than to those at higher elevations (Ficht and Heinton, 1939). Data showing this are given in table 84.

Table 84
European corn borer moths taken in traps at elevations
(data from Ficht and Heinton, 1939)

Height of trap (feet)	5	10	15
Moths (number)	913	549	331

It is interesting to note that Glick (1939) caught 225 lepidopteran adults within 5000 feet.

More pink bollworm moths, *Pectinophora gossypiella* (Saund.), by sexes were collected by light traps at two feet than at heights to fourteen feet. This is shown by data given by Glick *et al.* (1956) in table 85.

Table 85
Female and male pink bollworm moths at elevations
(data from Glick *et al.*, 1956)

Height (feet)	2	4	6	8	10	12	14
Collected at height (percentage)	39	16	16	8	8	7	6
Females collected (percentage)	23	17	20	17	12	14	14
Males collected (percentage)	77	83	80	83	80	86	86

A higher percentage of males than of females was taken at each height. Slight sex differences are also indicated, in which higher percentages of females than males were taken at lower heights.

Pink bollworm, *Pectinophora gossypiella* (Saunders), moths were collected by Glick (1967) at different altitudes. The number of moths reportedly taken in total minutes was reduced to the number of minutes required to collect one moth (table 86).

Table 86
Pink bollworm moths collected at altitudes
over Mexico and the United States
(data from Glick, 1967)

Altitude, (ft.)	20	100	200	500	1,000	2,000	3,000	4,000	5,000
No. min. to collect a moth	87.5	120.2	366.7	222.8	401.0	691.7	1,770.0	0	0

These comparative frequency data show that most moths were taken at 20 feet and became fewer at higher altitudes, to zeros at 4,000 and 5,000 feet.

Species differences indicative of vertical abundances were observed by Deane *et al.* (1953). *Anopheles darlingi* Root were more numerous in a forest at ground level, while *A. mediopunctatus* (Theo.) and *A. shannoni* Davis were least abundant at ground level and more numerous at higher levels, as shown in table 87.

Table 87

Anopheles of three species at elevations
(data from Deane *et al.*, 1953)

Height (meters)	0	5	10	15
Percentage of				
A. darlingi	51	18	19	12
A. mediopunctatus	3	6	29	62
A. shannoni	4	7	25	64

Malarial mosquitoes, *Anopheles quadrimaculatus* Say, are more abundant at ground level than at higher elevations, according to Goodwin (1942), as shown by data given in table 88.

Table 88

Malarial mosquitoes affected by elevations
(data from Goodwin, 1942)

Height	0	1.5	3.0	4.5	6.0	7.5
Malarial mosquitos (no.)	27	16	5	10	6	4

All species of mosquitoes were more numerous at lower levels in each of two locations, according to MacCreary (1941), as shown in table 89.

Table 89

Mosquitoes at different elevations
(data from MacCreary, 1941)

	Height (feet)		
Location	4.5	80.0	103.0
Gordon Heights	3166	------	447
Fenwick Island	14642	1374	-----

In catches by Horsfall (1942), 77 percent were at the four-foot level.

Frit flies, *Oscinella (Ocinis) frit* L., were reportedly most abundant at 16 feet, although they were also numerous between 33 and 60 feet. By use of adhesive paper, Korting (1931), discovered that frit flies were most numerous at ground level and decreased in numbers with elevation increases. He reported, however, that two thysanopterans, *Limothrips cerealium* Hal and *Haplothrips aculeatus* F., were most numerous above three feet.

Fruit fly (presumably *Drosophila* spp.) populations were most numerous at 0.54 and least abundant at 5.04 meters, according to Profft (1939). This is shown by the values in table 90.

Table 90
Fruit fly abundances as affected by elevations
(data from Profft, 1939)

Height (cm.)	5.4	12.4	20.4	27.9	35.4	42.9	50.4
Fruit flies (no.)	397	319	245	205	146	136	114

Samples indicating vertical dispersion of *Drosophila* spp. were taken by Basden and Harnden (1956) in the Arctic Circle. The range of elevation, although greater than that given by Profft (1939), suggests a more rapid rate of deceleration than that given by Profft (1930). Observations are given in table 91.

Table 91
Abundances of *Drosophila* spp. at elevations
(data from Basden and Harnden, 1956)

Height of traps, feet	10	20	30
Drosophilidae trapped, number	322	69	47

Vertical dispersion of small organisms is caused by the sum of all forces reacting successfully against gravity to move and to elevate individuals. Patterns of distribution are seen in the above data, in which gradients of a species near ground level become fewer and fewer with increasingly higher and higher altitudes. The differences are those of degree and may more properly be considered as relative abundances, than as "terrestrial," "plankton," or "aerial" zones. It is important to note the rate of decrease with altitude increase or the form of the gradient involved.

Data on vertical dispersion are sufficiently abundant to provide some understanding of the patterns of distribution. Semilogarithmic and a modified semilogarithmic formulae were used by Wolfenbarger (1946) to show the relationships between aerial densities and altitudes. These successfully depicted the form most gradients assumed. A more extended discussion of the relationships of populations at different heights, however, was given by Johnson (1957). The relationship he considered closer than others was

$$f(z) = C\,(z + z_o) - \lambda$$

These characters express as follows:

$f(z) =$ density at height z

$C =$ a scale factor depending on population size

$\lambda =$ an index of the diffusion process and of the profile

$z_e =$ a parameter whose significance reportedly depends on the exchange of organisms between air and ground

Close relationships of the observed and the calculated values of many sets of data were shown. The formula admittedly may not hold, however, during the initiation or the termination of a mass movement of organisms.

Altitudinal profile zonations. Three indefinite and indistinct zones or strata appear in a vertical profile of a population of dispersing organisms. Organisms initiate the dispersal journey at the lowest places and begin to rise essentially to disperse in a favorable medium. Such units may be in terms of fractions of millimeters or of many meters from solid or earth-bound materials. Increasing the elevation decreases the contact with barriers or objects that individuals might encounter in the dispersion journey. Organisms in the lowest strata are in the process of becoming "airborne"—provided they depart from the earth. This may be the threshold of the dispersion journey and be termed the "suboptimal" strata. Organisms move or may be moved to elevations where they avoid obstructions, to the economy of the species. The more favorable elevations, where most organisms are found, may be termed the "optimal" zone or strata. At altitudes above the optimal zone, numbers of organisms may be expected to decrease and approach zero. This may be considered the "supraoptimal" zone of the vertical abundance profile. Figure 26 is given to illustrate these zones.

A discussion somewhat similar to that above was given by Gregory (1962). It also indicates our paucity of information on vertical dispersion. In most cases the suboptimal zone, or stratum, is of shorter height than the other zones. For most organisms the optimal zone is a wider band than the sub-optimal zone. A much wider zone or stratum appears evident above the optimal peak of abundance, the supra-optimal band, whose maximum height is unknown. Life may cease to exist in the zone, so that organisms become inert organic materials.

Species or kinds of organisms doubtless vary widely in abundance in different heights. Such abundances doubtless tend to "peak," or assume mean averages, but to possess all manners of variations. Many factors affect the vertical profile.

Hill slope. A gravitational advantage or disadvantage may follow the initiation of the dispersal of an organism on a hill slope. Dispersion could be enhanced or retarded, depending on uphill or downhill movement. Data have been gathered on the incidence of white pine blister rust, a passive disperser organism, and on dispersion of a shield bug, an active disperser organism.

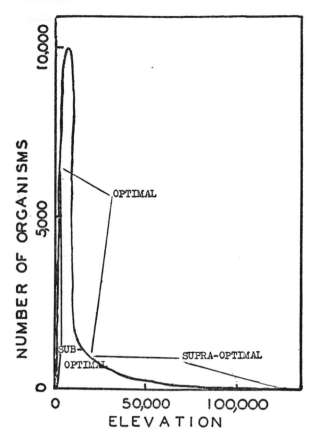

Fig. 26.   Optimal, sub- and supraoptimal profiles of vertical dispersion zones of organisms, a schematic diagram.

Incidence patterns (determined by cankers) of white pine blister rust disease, caused by *Cronartium ribicola* Fisch. were reported by Buchanan and Kimmey (1938) around the origin of disease organisms. Although the disease ordinarily assumed "comparable proportions in all directions," the authors found that more cankers were below rather than above the spore source. Numbers of *Ribes* spp. trees with one or more cankers were taken from the authors' figure 7 both uphill and downhill from a spore source. Two regression curves were drawn to show the elevation responses (fig. 27).
Both curves agree very closely with the observed values. Both curves differ widely initially but begin to converge to reach the same low incidence rates between 360 and 450 feet from the central *Ribes* spp. spore source.

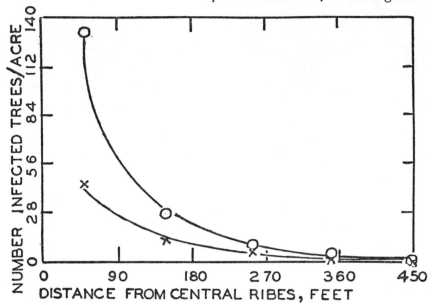

Fig. 27. Cankers of white pine blister rust disease on trees below (upper curve) and above (lower curve) the spore source (data from Buchanan and Kimmey, 1938).

Some data on hill slope effects on the dispersion of a shield bug, *Aelia rostrata* Boh., were reported by Brown (1965). Results as numbers and percentages of shield bugs are given in table 92.

Table 92
Hill slope effects shield bug abundances
(data from Brown, 1965)

Flight	No.	Direction in relation to hill slope		
		Uphill	Downhill	Across
Initiation all records	No.	219	388	208
	%	26.9	47.6	25.5
Wind, 6 ft./No.		58	132	31
Sec. +	%	26.2	58.9	14.0
Termination all records	No.	321	382	219
	%	39.1	34.3	26.6
Wind, 6 ft./No.		86	114	21
Sec. +	%	38.9	51.6	9.5

A tendency is seen for more downhill movement of these bugs than for uphill dispersal. Terminal dispersion was apparently influenced less by stimuli to disperse downhill than were initial movements. Percentages of downhill movements were increased by winds of 6 ft./sec. and over, and they were decreased by cross-slope influences.

Meteorological Factors

Various climatic factors, such as temperature, moisture, light and air movement, affect the distances to which organisms disperse or are dispersed. Many authors have written of the effects of atmospheric phenomena, although Cammack (1958) says that seldom have these effects been measured, especially with regard to distance. Considerable efforts are involved in securing definitive data with references to climatological factors affecting dispersion distance, although the results of these efforts occasionally leave more to be desired.

In addition to various specific climatological factors affecting the dispersion of organisms, there are many degrees of relativeness and of interrelationships of these factors associated with organisms. Light and darkness, for example, are cyclically prevalent and cannot be controlled under field conditions. Certain amounts of moisture are present with any and all temperatures, although relatively more moisture and lower temperatures are prevalent during hours of darkness. Climatological factors are present in degrees different from all other factors and need to be understood. The dispersion of organisms under field conditions is the net result in total movement resulting from all factors acting and reacting. Some assessments have been made of different climatic factors under field conditions; these assessments have value and are of interest in understanding the dispersion of organisms. Studies of dispersion under controlled conditions of the laboratory are also desirable and are advocated.

Although wind direction may be one factor, other factors determine the direction of flight of migrating painted lady butterflies, *Vanessa cardui* L., according to Abbott (1950). Of recorded observations, two-thirds of the flights were against the wind. There was no relationship between flight and angles of wind direction. There was an absence of directional effects in 470 recorded migrations of the painted lady butterfly, according to Williams (1958).

Great variability in the climate of any spot is recognized in the enormous "numbers of microclimates" near ground level, according to Geiger (1950). These are induced principally by variations in temperature, moisture, and wind and light. A small portion of these variations may be attributed to protected enclosures, to surroundings, and to laminary strata of air occurring about the perimeters or margins of all earth particles or objects. Those molecules near to and contacting the earth and its objects, although capable of being energized and moved, resist changes in position. More resistance to movement and to change is exercised by enclosure and by those molecular layers nearest the objects; less resistance to movement is exercised by those distant from the objects.

The atmosphere is a fluid restricted by gravity, heated by energy from the sun, and containing moisture that is turned to clouds, rain, and ice. There is much interplay of various factors in the atmosphere. This interplay makes many eddies or turbulences and creates as many differences, making the fluid a constantly moving, ever changing medium, according to Long (1960). A correct and theoretically functioning mathematical atmospheric diffusion model could be developed by Sutton (1947) only be admitting extreme values for "gustiness" in the dispersal of gases. This may serve to indicate some problems connected with wind dispersal of organisms in the lower atmosphere and the difficulty in determining a comprehensive theory. Air mass invasions are at the mercy of the underlying medium, especially nearer the earth's surface. This may be further understood by Sutton's work (1947) showing how gas clouds are distorted by surface objects, even by ordinary vegetation. Turbulences (discussed below and related to the earth and objects present on it) also assist in creating great inequities in climatic factors, thus affecting the distance to which dispersal of organisms occurs. Air is subject to much overturning and mixing, in which the fluid motions involved comprise one of the most unstable and complex of the physical sciences.

Earlier reviews on the relationship of weather to fungus and bacterial diseases were given by Foister (1935, 1946), Gregory (1945), and Stakman and Christensen (1946). An earlier review of insects and climate by Uvarov (1931) devoted some discussion to insect dispersion, as did the works by Glick (1939), Freeman (1945), and Wellington (1945, 1945a, 1945b, 1945c, 1954). Many treatises that provide qualitative discussion of climate and weather in connection with dispersion of small organisms will be omitted here. Although such references are valuable, it is the definitive or *quantitative* dispersal distance measurements that are significant for this study.

Most insects apparently tend to maintain control over their dispersal distance, dispersal direction and dispersal time. Such control is presumably for the economy of the species. Although instances of distant insect movement during hurricane winds may occur, such instances seem unusual and uncertain. Insects generally seek protection from high winds in order to disperse later according to those stimuli that motivate them. Infestations of broad mite, *Hemitarsonemus latus* Banks, were observed by the author subsequent to a hurricane in which it was believed that many mites had been blown or washed away from lime fruits and were doubtless lost. Many individuals remained, however, and were often near and between fruit calyces, where they could have been largely and protectively enclosed.

LIGHT OR DARKNESS

Light is practically inseparable from other climatic factors in determining effects on the dispersion of small organisms. Consideration is given, however, to light (or darkness) where possible.

Sporangia of the potato late blight disease organism, *Phytophthora infestans* (Mont.) de Bary, were found dispersing between 8:00 A.M. and 6:00 P.M., with peak concentrations between 9:00 A.M. and 12:00 noon, according to Bawden (1951). Whether such hours are favorable to the organism or its host plants and whether this situation is attributable to temperature, moisture, light, and/or rhythm affecting the fungus or host plant, or to some other factor or combination of factors, remains to be determined.

Uredospores of wheat stem rust, *Puccinia graminis* Pers., are in the disperse phase about midday, according to Bromfield *et al.* (1959). Data from one sampling station (#9) of spore counts taken on coated vertical rods over a six-day period are summarized in table 93.

Table 93

Wheat stem rust spores at hours sampled
(data from Bromfield *et al.*, 1959)

Hours of sampling period	*700-1000*	*1100-1500*	*1500-1900*
Avg. No. spores/day	101	253	62

Rye (*Secale cereale* L.), cocksfoot (*Dactylus glomerata* L.), ryegrass (*Lolium* sp.), timothy (*Phleum pratense* L.), and sugarbeet (*Beta vulgaris* L.) pollen dispersions were related to daylight hours, according to Jensen and Bøgh (1941). Rye pollen disperses throughout the entire day, while other species disperse actively at the hours listed in percent, in table 94.

Table 94
Pollen abundance of four grass species and
of the sugarbeet as affected by hourly collections
(data from Jensen and Bogh, 1941)

Plant				*Hours*				
	5-6	*6-7*	*7-8*	*8-9*	*9-10*	*10-11*	*11-12*	*12-13*
Rye	10	36	41	39	25	33	46	78
	8	41	43	100	12	11	89	21
Cocksfoot	—	—	11	100	55	18	—	—
Ryegrass	—	—	—	—	10	100	17	88
Timothy	—	100	45	18	—	—	—	—
	9	43	100	6	—	—	—	—
	28	36	83	100	—	—	—	—
Sugarbeet	—	—	100	70	37	18	29	5

Plant		*Hours*			
	13-14	*14-15*	*15-16*	*16-17*	*17-18*
Rye	97	100	65	32	23
	37	42	47	4	4
Cocksfoot	—	—	—	—	—
Ryegrass	—	—	—	—	—
Timothy	—	—	—	—	—
	—	—	—	—	—
	—	—	—	—	—
Sugarbeet	—	—	—	—	—

Wind as a climatic factor was related to pollen dispersal in some of the above data and may have accounted for the relative abundances or scarcities during certain of the above indicated hours.

Rempe (1937) found that tree pollen of various species dispersed more abundantly in day than in nighttime hours, at altitudes of from 10-40 to 1500 feet, table 95).

Table 95
Tree pollen abundance at elevations by day and night collections
(data from Rempe, 1937)

Altitude (meters)	10-40	200	500	1,000	1,500	Totals
Day flights	904	849	852	581	267	3,453
Night flights	557	560	283	85	45	1,550

It is seen, therefore, that more than twice as much pollen was taken in day than in nighttime hours. A more important difference is that correspondingly more pollen was taken at the higher altitudes in daytime than in nighttime hours. This suggests that a sedimentation process was more active, or at least more effective, at night. Fewer updrafts during the hours of darkness may hasten sedimentation processes at night.

Atmospheric phenomena influence pollen shedding, and there are likely to be close relationships between the time of pollen dispersal and certain climatic factors. Although inflorescences of various species of Gramineae open for a week or more of pollen shedding, Jones and Newell (1946) found that the time of day in which the blooming occurs is often restricted to a few hours. Their data, obtained from exposed slides, shows definite hours in which pollen is dispersed (table 96).

Table 96
Hourly dispersion of some grass pollens
(data from Jones and Newell, 1946)

Species	*Avg. days of pollen dispersal*	*Hours of pollen dispersal*
Agropyron cristatum (L.) Gaertn., crested wheatgrass	8	2-6 P.M.
Agropyron intermedium (Host) Beauv., intermediate wheatgrass	10	2-6 P.M.
Agropyron smithii Rydn., western wheatgrass	7	2-6 P.M.
Andropogon furcatus Muhl.	7	4-7 A.M.
Bouteloua gracilis (H. B. K.) Lag., blue grama	6	3-9 A.M.
curtipendula (Michx) **Torr.**	8	4-9 A.M.
Bromus inermis Leyss., bromegrass	10	**2-7 P.M.**
Buchloe cactyloides (Nutt.) Engelm., buffalograss	Indeterminate	6-11 A.M.
Festuca elatior var. *arundinacea* (Schreb.) Wimm. tall fescue	9	1-6 P.M.
Panicum virgatum L., switchgrass	12	10 A.M.-2 P.M.
Poa pratensis L., Kentucky bluegrass	8	3-8 A.M.
Secale cereale L., rye	8	3-11 P.M.
Zea mays L., corn	10	7 A.M.-4 P.M.

Evans (1951) found that the behavior of dispersing *Lepidochitona cinareus* was affected by light. Movement was reportedly greater in artificial or daylight than in darkness. Some data were given, however, for movement in darkness after a ten-minute waiting time (table 97).

Table 97
Chiton dispersal in darkness
(data from Evans, 1951)

Cm. from release point	0	0-3	3-6	9	12
Number of chitons	14	2	1	2	1

Since most animals remained at the zero distance, it may be seen that little dispersal occurred in the time of darkness.

More studies have been made, apparently, about the effect of climatological factors on aphid dispersion than on any other group of insects, partly because they transmit viruses that are of much economic importance (Johnson, 1954).

Diurnal aphid dispersion activities were recognized by Johnson (1954, 1957) and Taylor (1958), (fig. 28 and 29). Diurnal fluctuations

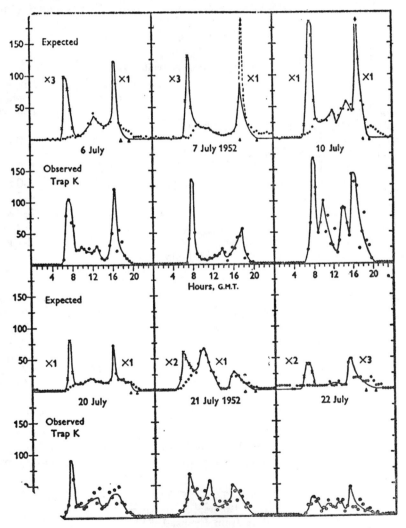

Fig. 28. Diurnal flight curves of aphid abundance (fig. 2 from Johnson *et al.*, 1957).

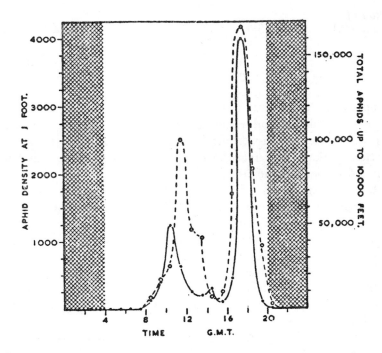

Fig. 29. Diurnal flight curve of aphid abundance. Solid line is aphid density per 10,000,000 cubic feet of air at vegetation level. Dash line is the total aphid content of a column of air of 10,000,000 square feet basal area from 1 to 10,000 feet altitude (fig. 1 from Taylor).

occur and tend to have two peaks of aphid abundance. One is a morning hour peak, which tends to occur about 10:00 A.M. The peak of greatest abundance, however, tends to occur about 14:30; although there is considerable variability in the peak of abundance. Aphids do not fly at night, according to Broadbent (1949).

It is observed in the above figures that the data were taken at vegetation level. The insects collected at higher elevations cover a wider spread of time than those at lower elevations or near the source of the aphids. This is shown in table 98 by data from Johnson (1951).

Table 98

Differentials in aphid abundances at elevations

(data from Johnson, 1951)

Elevation of collection (ft.)	2:30	5:30	8:30	11:30	14:30	17:30	20:30	23:30
50	0	0	0	41	117	11	0	0
500	0	0	0	15	13	4	4	1
1000	2	0	0	0	1	0	0	4

More aphids were taken at lower than at higher elevations, and they were taken within a shorter period than at higher elevations. Fewer aphids were taken at higher elevations, and these were spread over wider time periods. This may be of value to the insects in that more widespread dispersal is achieved.

Only daylight flights of the migratory grasshopper, *Melanoplus mexicanus mexicanus* (Sauss.), were observed by Parker *et al.* (1955). Initiation of flights usually occurred over a two-hour period between 11:00 A.M. and 1:00 P.M. Termination of flights always occurred several hours before sunset. Flights were terminated also during the day whenever clouds obscured the sun and reduced light intensities. Light intensity is one controlling factor of migratory dispersal flights; temperature is the other, and there may be a relationship between light intensity and temperature.

More cotton boll weevils, *Anthonomous grandis* Boh., a nonswarming species, were in flight in morning than in afternoon hours, according to Fenton and Dunnam (1928). The flight time of a portion (six percent) of trapped weevils was not determined.

Larvae of the gypsy moth, *Porthetris dispar* L., exhibited two daily periods of dispersion, 9-12 A.M. and 2-5 P.M. (Minott, 1922). These larvae in search of food are dispersed by wind to nearby plants, and this is one of the dispersal phases of the species.

Definite diurnal cycles of *Drosophila pseudoobscura* Frowola involve periods of peak activity in early morning and late afternoon according to Mitchell and Epling (1951). Such activities are related to sunrise and sunset on clear warm days and to local light intensity. Shade by clouds and mountains restricts dispersion activities and alter the hourly span and the number of flights. Restrictions imposed by light intensity may be temporary or prolonged. Temporary restrictions reduce movements over a short lifespan. Prolonged restrictions may reduce total seasonal movement.

The tsetse fly, *Glossina morsitans* Wst., was found dispersing in a thicket (partial shade) more than in an unshaded area. Although temperature and moisture may be factors affecting the distance of dispersal of the fly following its host animals, light may be more important. Flights into thickets and into clearings, seeking host animals, show entirely different rates of dispersal. Data from Swynnerton (1936) illustrate these differences in table 99.

Table 99

The effect of ground coverings on the distances to which tsetse flies disperse
(data from Swynnerton, 1936)

Distances of flight following host (yds.)	50	100	200	300	500	1,000	2,000	4,000	6,000
Flights into clearing	—	12	4	1	—	—	—	—	—
Flights into thicket	27	—	—	—	19	18	10	2	1

Males of the honeybee, *Apis mellifera* L., returned to the colony practically equally, whether they were released in the ante- or postmeridian portions of the day, according to Oertel (1956). Honeybee colonies nearer foraging crop plants continued evening flights later than did colonies more distant from the crop. Earlier morning flights from nearby colonies were believed more likely than from more distant colonies (Ribbands, 1951). Counts of returning honeybees, per five-minute period, taken on each of two days and from each of two locations, were combined and show the extent of approaching darkness, as in table 100.

Table 100

Twilight hours affecting numbers of returning honeybees at distances from crop plants
(data from Ribbands, 1951)

Greenwich mean time, P.M.	Distance from crop (miles)		
	0	⅜	¾
8:30	1,776	1,464	551
8:45	615	240	62
9:00	131	15	4
9:15	80	0	0

Reductions in the numbers of returning bees were observed from all distance points. More rapid reductions, however, appear to result from longer distances.

MOISTURE, WETNESS AND DRYNESS

Relative amounts of moisture are required for the dispersion of organisms. Dessication is an important factor in the destruction of many dispersing organisms. On the contrary, rains precipitate many spores, pollen, and other organisms and thus terminate the dispersion of many individuals. Owing to the changes of weather, moisture extremes (wetness and dryness) frequently occur and are encountered in the course of the dispersion process. Such extremes are hazardous and are undoubtedly most frequently encountered by those species that spend the most hours and cover the longest distances in the disperse phase. Presumably those species that spend much time (several days) and move long distances (hundreds of miles) in the dispersion process are able to adapt themselves more effectively to wider ranges of moisture or dryness. Although many qualitative statements are present throughout botanical and zoological literature, definitive data on the effects of moisture or dryness on the distance of dispersal of organisms are very meager, as is shown by various reviews: Foister (1935, 1946), Freeman (1938), Glick (1939), Gregory (1945), Horsfall *et al.* (1960), Ingold (1939), Proctor (1934, 1935), Stakman and Christensen (1946), Wolfenbarger (1945, 1959).

Rainfall must affect atmospheric spora in several ways. Widespread distribution of rainfall is probably more important than quantity over a given period. Rainfall probably (1) increases spore germination, colonization, growth and production; (2) washes spores out of the air, decreases spore concentration, reduces widespread dispersal; and (3) washes spores off conidiophores, reducing the number that might become airborne.

Wind-blown rain was believed by Faulwetter (1917) to be an important agent of transmission for disease organisms. He conducted experiments and reported extensively on some phases of water droplet spatter. Spatter distance was found to vary considerably, depending on volume, size of drop or volume, distance of fall before impact, and whether the drop fell on glass or on blotting paper. Based on drop size or volume, a summary of the data combining fall and all heights (one, two, and four feet) is given in table 101.

Table 101
Splash distances of drop sizes
(data from Faulwetter, 1917)

Drop size (ml.)	0.10	0.06	0.05	0.02
Splash distance (in.)	17.5	15.3	13.3	12.1

Spatterings from the larger drop sizes disperse farther than those from the smaller drops. (Further reference is made to this factor and to the principle involved under Density Levels at the Origin in chapter 7.)

Energy supplied by waterfall from greater heights disperses spatter droplets to longer distances than does waterfall from lesser heights. This is shown in a summary of data from Faulwetter (1917) combining all drop sizes (0.10, 0.06, 0.04, and 0.02. ml.), listed by fall distances in table 102.

Table 102
Splash distances of height of fall
(data from Faulwetter, 1917)

Fall distance (ft.)	1	2	4
Splash distance (in.)	11.3	14.9	17.5

Linear relationships of the above factors appear from plotting the above data after logarithmic transformations.

Influence of the receptor of waterfalls on distance of water splashes was found by Faulwetter (1917a) to produce differences. Droplets dispersed farther following impacts on comparatively soft and spongy blotting paper than on dense, hard glass surfaces. Data were given on splash distances for drop falls of one, two, four, and sixteen feet for the different drop sizes. A summarization of the data combining heights fallen for each drop size is given as maximum spatter, in inches, in table 103.

Table 103
Effects of fall receptors of drop sizes
(data from Faulwetter, 1917a)

Receptor	Drop size (ml.)				Total
	0.10	0.06	0.04	0.02	
Blotting paper	210	180	148	138	676
Glass	186	142	122	100	550

An explanation for the longer distance of splash from impacts on blotting paper than from impacts on glass may be in the relative amounts of surface tension. There appears to be nothing in either receptor surface that would increase the energy over that of the distance of fall. Some factor, therefore, may reduce the energy supplied by the fall. Glass surfaces may possess more surface tension than blotting paper. The means by which wider splash occurs from blotting paper than from glass, however, is not clear. The size of the droplet, although not discussed, may result from the impact on different target materials and influence the distance of splash.

Larger splash drops were found by Faulwetter (1917a) at longer distances from the point of impact. The explanation was that the smaller drops encountered more resistance to air than larger drops. Ratios of weight to the areas exposed, it was pointed out, were less with small drops than with large drops.

The dynamics of water splash as an agent in dispersing micro-organisms vary with different conditions, as shown by Gregory *et al.* (1959). Estimated kinetic energy (ergs) for different drop diameters and for different drop diameters, are given in table 104.

Table 104

Water splash affected by drop size and height of fall
(data from Gregory *et al.*, 1959)

Height of fall (m.)	*Actual drop diameter (mm.)*		
	2.9-3.0	3.8-4.0	4.9-5.0
2.9	1950	5900	14300
7.4	4200	11800	24800

A considerable increase in kinetic energy (ergs) is seen in the 7.4 m. (higher) fall over the 2.9 m. (lower) fall.

Surface films placed in the impact or target areas of the falling water drops were used to indicate spatter (Gregory *et al.* 1959). These films were found to affect the number and size of droplets produced by falling drops. Data in evidence of the droplets produced are given in table 105.

Table 105

Surface film thickness and composition
affecting number of droplets splashed
(data from Gregory *et al.*, 1959)

Surface film thickness	*Film composition*	*Total no. of droplets*
No film	Dry glass	27
0.1 mm..	Containing spores	1055
0.2 mm.	Containing spores	868
1.0 mm.	Containing spores	462
0.1 mm.	Tap water	1085

Deeper films tend to absorb fall energy without splash, as shown by the inverse relationships of film thickness and droplets produced. The scarcity of droplets produced from falling on dry glass suggests that surface tension retains the falling water and that the glass absorbs the fall energy. The threshold of increase from no film to film 0.1 mm.

thick was not determined, owing to the difficulty of producing very thin films. Although 30 more droplets dispersed from tap water than from water containing spores, the proportion is comparatively small. The distance of dispersal of the droplets was also affected by different film thickness. More droplets splashed further as a result of falling on 0.1 mm. film thickness than on 0.5 mm. and on 1.0 mm. thicknesses, as shown in fig. 30.

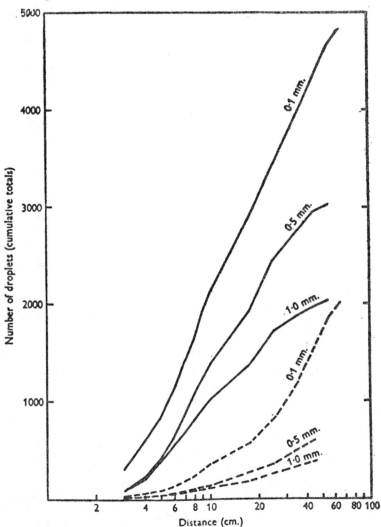

Fig. 30. Spore and water drop dispersal as related to distance and to target film thickness. Solid lines are water droplets. Dash lines are droplets with spores (data from Gregory *et al.*).

Organisms may be borne by water in dripping from plants, or they may be present on the surface water and dispersed by the dropping of water from plants. Spores are dispersed whether they are in the incident drop or are in the surface film. This is shown by data from Gregory *et al.* (1959) in table 106.

Table 106
Target film effects with spore suspension
(data from Gregory *et al.*, 1959)

Target film thickness (mm.)		Target film is spore suspension; incident drop is tap water	Target film is tap water; incident drop is spore suspension
0.1	No. of droplets	876	886
	No. with spores	200	261
0.1	No. of droplets	462	520
	No. with spores	118	192
Totals	No. of droplets	1338	1406
	No. with spores	318	453
Percentage with spores		24	32

More droplets, more spores, and a slightly higher percentage of droplets contained spores in those occurrences in which the incident drop was a spore suspension and the target film was tap water.

Horizontal distances of droplet dispersal in still air extended to about 100 cm., with droplets falling nearest the incident drop fall site. Tabulation of data (Gregory *et al.*, 1959) for drop diameters of 3, 4, and 5 mm. for 2.9 and 7.4 m. heights of fall are given in number of droplets deposited per sq. cm. in one splash for horizontal distances. These are given in table 107.

Table 107
**Dynamics of water droplet size and height of fall
on water droplet dispersal as affected by distance**
(data from Gregory *et al.*, 1959)

Distance from target (cm.)	2-3	2-5	6-7	8-9	10-18	25-35	35-45	45-55	55-65	65-75
Drop diameter (mm.)				*Height of fall, 2.9 m.*						
3	3.75	2.79	0.96	1.03	0.28	0.04	—	—	—	—
4	7.45	5.14	2.06	1.26	0.38	0.08	0.20	—	—	—
5	12.30	6.54	5.18	2.46	0.78	0.36	0.12	0.16	0.04	—
				Height of fall, 7.4 m.						
3	2.74	1.98	1.26	0.86	0.23	0.06	0.04	0.02	—	—
4	19.20	8.23	7.22	5.27	1.20	0.28	0.16	0.10	0.04	—
5	14.80	9.81	8.92	6.93	1.30	0.14	0.12	0.20	0.08	0.04

Most droplets fall nearest the point of impact and become fewer in number as the distance from their origin increases. Gradients so indicated occur regardless of height of fall or of drop diameter. Droplets disperse farther from larger drops (higher populations at the origin) and from longer distance falls than from smaller drops and shorter distance falls. Spore-carrying droplets were most frequent at the greatest distances of splash, 45 to 75 cm. This is shown in table 108.

Table 108
**Percentage of droplets with spores
at distances from splash point origin
(data from Gregory *et al.*, 1959)**

Distance of splash (cm.)	*Droplets with spores (%)*
2-3	25
3-4	23
4-5	21
5-6	19
6-7	17
7-8	13
8-9	22
9-10	30
10-18	42
18-25	53
25-35	43
35-45	67
45-55	100
55-65	100
65-75	100

There was a steady decrease in the percentages of droplets with spores from 2-3 to 7-8 cm. droplet dispersal and an increase from 8-9 to 45-55 cm. There is no known reason for this phenomenon.

There was an increase also in the percentage of spore-carrying droplets as droplet diameters increased in size. This is shown by data from Gregory *et al.* (1959) in table 109-110.

Table 109-110
Droplets with spores based on diameter of splashed droplets
(data from Gregory *et al.*, 1959)

Diameter of droplets	Droplets wtih spores (%)
15	0
16-21	1
22-27	3
28-33	3
34-39	11
40-45	13
46-57	10
58-82	29
83-116	18
117-164	51
165-231	89
232-328	95
329-464	54
465-655	90
655	100

Occasional reversals are seen in which droplets of smaller size contained larger percentages of conidia than droplets of larger size; this may be attributable to some factor in sampling that is without especial significance.

Soil moisture affected the emergence and initial dispersal of the beet eelworm, *Heterodera schactii* Schmidt, according to laboratory tests by Wallace (1955). Exposure of 100 cysts at 25°C placed in sand at different water-pressure deficiencies for 7 days, followed by determination of the distance of dispersal of individuals, was the method of experimentation. Results are given as the average numbers of nematodes at various distances in table 111.

Table 111
Dispersal of the beet eelworm based on water pressure deficiency
(data from Wallace, 1955)

Water pressure deficiency (cm.)	*Cm. from release site*			
	0-0.75	0.75-2.00	2.00-3.00	3.00-4.00
6-8	70.0	20.8	6.9	2.4
14-16	64.7	23.2	9.0	3.7
22-24	63.9	20.6	9.4	6.2
30-34	80.8	13.3	6.0	0
50-54	93.0	7.0	0	0

The results suggest that at water pressure deficiencies of more than 25 cm. there is reduced larval dispersion. At the lower, 6-24 cm. pressure, there was more dispersal in excess of 0.75 cm. distance from the release site than from 30 or more cm. pressure. Summer crimp nematodes, *Aphelenchoides besseyi* Christie, move over the surface of plants wet with dew or rain, permitting more widespread dispersal, according to Christie (1959). Data were not given, however, showing the distance of movement, as was done in regard to saturation deficiency by Wallace (1955).

A lower percentage of the red spider mite, *Tetranychus urticae* Koch, dispersed at lower than at high humidities, according to Hussey and Parr (1963). Data are given to show this in table 112.

Table 112
Red spider mite dispersal at humidities
(data from Hussey and Parr, 1963)

Relative humidity (%)	100	92	77	70
Mites that dispersed (%)	11	25	80	95

Lower relative humidities were shown to have motivated mite dispersal.

After studying vertical abundances of a rather wide variety of organisms, Vinje and Vinje (1955) concluded that "the higher the relative humidity at ground level the lower the number of objects recovered from the air."

Air pollutants, increasing with the expansion and multiplication of populations and of industrial processed, are accelerating from various enterprises. Damages from industrial processes are apparently increasing. Peach foliage injury, defoliation, and early fruit drop, for example, "was more pronounced following wet periods of low wind velocity," according to Daines *et al.* (1960). These observations are congruent with the theoretical considerations of Bosanquet and Pearson (1936), which hypothesized that larger amounts of acid and dust are precipitated by rains than during dry periods. Pollutants in the dispersion process apparently adhere to or enter into solution with raindrops and moisture particles and fall nearer the source of the pollutants than in dry periods. Further evidence of the distance effects of pollutants was seen in the findings by Daines *et al.* (1906) that injury "is usually restricted to an area" near the source of the emission. No definitive data were given, however, to illustrate the response by plants at different distances.

Black arm disease of cotton, caused by the bacterium *Pseudomonas malvacearum* EFS, was found concentrated on fields nearest old cotton land, according to Andrews (1936). Rains were believed to spread the organism over a distance range expressed in table 113.

Table 113

Black arm disease of cotton as related to old cotton fields
(data from Andrews, 1936)

Distance from old cotton fields (m.)	1	2	105	176	280
Plants diseased (%)	55	40	10	6	2

Maximum distance of disease incidence was in excess of 280 meters. Rains, as surface water, and not raindrop splash, were believed responsible for carrying plant debris containing spores of the organism.

Conidia of a head blight disease, *Gibberella saubinetti* OUD, are mucilaginously coated and were believed by Ishii and Koyama (1952) to be dispersed by water drops. Wind was believed not able to separate these individual spores from aggregate groups of spore masses.

Wet soil or mud, adhering to animals, autos, and other moving objects, disperse seed and other organisms as dispersal agencies (Etter, 1948). Although organisms may be dispersed by dry agencies, some species may be moved farther with massed wet soil. Obviously, the adhesiveness of wetted soil particles permits better attachment to moving objects. Such soil particles later dry and may fall in a new location, taking along a small amount of medium from the origin. More movement of soil and soil particles may be expected along roadways, according to Clifford (1959). More mud and more organisms will be dropped nearest the origin and increasingly fewer with increased distance from origin, although no data are known that show this phenomenon. Small seeds have a better chance of dispersal in mud than do large seeds, according to Clifford (1959).

Humidity. Relative humidity affected the spore takeoff of *Piptocephalus virginiana* Ledbeater and Mercer, according to Zoberi (1969), at wind speeds of 3.6 m. sec and at the relative humidities given in table 114.

Table 114

Initiation of spore dispersal at relative humidities
(data from Zorberi, 1961)

Relative humidity	52.5% (15.5°C.)	90.0% (16.0°C.)
Avg. no. spores/min.	2221	343

Spore takeoff was more abundant at 53.5% than at 90.0% relative humidity (if it may be assumed that the 0.5°C. temperature difference has no significance).

Movement of eelworm (*Heterodera schactii* Schmidt) larvae were influenced by soil moisture (Wallace, 1955). Results suggested that at pressure deficiencies of more than about 25 cm. of water, larval mortality is reduced, while at low deficiencies (6-8 cm.) dispersion is more widespread.

Owing to the variability of relative humidity and its changes with changing temperature and to the lack of relationship of insects collected at relative humidities, this method of classification was not used by Glick (1939). Collections of insects at higher altitudes (200-5000 ft.) showed a relationship of insect abundance with the dew point at ground surface. Data from the graphic figure (No. 6) for a tabular presentation are approximate and the 3,000 and 5,000 feet classes are omitted, since all were less than one insect. A summarization is given in table 115.

Table 115
Insects collected at elevations
and at ground surface dew points
(data from Glick, 1939)

Surface dew point (F°)	27	32	37	42	47	52	57	62	67	72	77
Altitude (ft.)											
200	4	7	11	12	12	13	15	17	15	14	11
1000	1	2	4	5	5	5	6	6	5	5	4
2000	1	1	2	2	2	3	3	3	3	3	2

Peaks of abundance were at or near 62° F. with 200 and 1,000 foot altitudes, and wide peaks were seen at from about 37° to 72° for all collections. A relationship between the number of insects taken and the actual quantity of water vapor per volume of air (absolute humidity) was shown by Glick (1939) in his Figure 8, from which the data in table 116 are taken.

Table 116
Insects collected by night and day flights at vapor pressures
(data from Glick, 1939)

(mid-class points)

Vapor pressure (in.)	0.150	0.250	0.350	0.450	0.550	0.650	0.750	0.850	0.950
Day collection	4	8	9	10	12	10	11	8	7
Night collection	—	6	11	12	13	15	11	8	7

These data are average numbers of insects per 10-minute flying time at altitudes of 1,000 feet or less. Since temperatures were operative, according to Glick (1939), there is a question of the extent of their effects in the above measurements.

Collections of insects by Freeman (1945) showed that more insects were taken at the lowest relative humidity range and fewer insects at higher humidities (table 117).

Table 117

Insect collections at heights and relative humidities
(data from Freeman, 1945)

Humidity (% R. H.)		37-53	54-59	60-64	65-73
Height of collection (ft.)	177	70	26	44	29
	277	36	21	26	16
Totals		106	47	70	45

Higher humidities increased the aggregation of the white pine cone weevil, *Conophthurus coniperda* (Schwarz), according to Henson (1961). Average indexes of two "runs" of each of two relative humidities are given in table 118.

Table 118

Aggregation of white pine cone weevils at relative humidities
(data from Henson, 1961)

Relative humidity	40%	100%
Average index*	0.315	0.270

*A complete index is 0; that of random association is 1. Statistically significant differences, indicating two different populations of grouping insects, were obtained from the data.

An extreme in the need of proximity of water for organisms is seen in the human habitation-tsetse fly habitat in Africa. Natives must have water; hence they live near streams, in compounds. Tsetse flies, *Glossina* sp., also live along the streams in fly belts. Those compounds in and nearest to fly belts suffer most from the sleeping sickness disease, caused by a *Trypanosomata* organism, transmitted by the flies. People in many compounds perished from the disease; other people moved farther from the streams. It was necessary to carry water from the permanent streams, however, which limited and even reduced distances to which compounds were established. Ruined compounds were found by Morris (1952) to be frequent within a mile of fly belts and less frequent with increasing distances from tsetse fly habitats along rivers or other streams. The average incidence of ruined compounds is given in table 119.

Table 119
Incidence of ruined compounds at distances from ruins
(data from Morris, 1952)

Miles from river	0.13	0.38	0.63	0.88	1.25	2.00	3.00	4.00
Ruined compounds (%)	82	52	45	34	23	18	17	17

Depopulation of compounds was reduced by almost 6/7 over a distance of four miles, according to a regression computed from these data (Wolfenbarger, 1959). Zero percentage of ruined compounds was reached at some distance in excess of four miles, the maximum distance of the observations.

Less spectacular than its effect on determining human habitations is the habit of the tsetse fly, *Glossina palpides* Newst., to follow host animals more often during a wet than in a dry season. Although the fly apparently attempts to remain in shade or semishade, moisture is doubtless an important factor. Flies do, however, follow host animals (man) from thickets in which they rest. This is indicated by data from Moggridge (1949) (table 120).

Table 120
Wet and dry season incidences of tsetse fly
following host animals
(data from Moggridge, 1949)

Yards from thicket margin	25	50	80	100	125	180	230	280
Flies during wet season	—	—	35	—	—	12	10	2
Flies during dry season	53	13	—	7	2	—	—	—

Two very different sets of data are observed, in which distances more than two-fold greater were traversed in the wet than in the dry season. Further reference is made to moisture under "Barriers," chapter 6.

Proximity to a lake or river induced a higher incidence of hypersensitivity to emanations of a caddis fly (Trichoptera) (Parlato *et al.*, 1934). A distance of six miles from the river provided almost perfect relief from the emanations.

It is difficult, or often impossible, to ascertain effects that are entirely attributable to a single factor such as moisture. In the above instance the emanations may have originated entirely from the caddis flies along the river; this presents a debatable instance of the effects of wetness.

Rainfall affects the dispersal of organisms in several ways during the initiation, course, and termination of dispersal. An adaptation from

Richards (1956) is given to show actions subsequent to rainfall; this explanation applies to organisms in general, as follows:

> Rainfall may initiate the dispersal of many seeds, spores, insects, and individuals of other species and increase the length of time of viability. Sedimentation of many or even of most disseminules terminated the movement of many individuals and shortens distances following precipitation. Heavy or prolonged rains precipitate and destroy many individuals. Drought (absence of rain) may reduce plant growth and consequently reduce the amount of host plant or food material available. Such reductions may increase the distance dispersed by organisms by providing more favorable living conditions. On the contrary, some species may disperse shorter distances during drought periods.

Sunlight may aid in destruction of life of dispersing organisms, especially those long exposed, thus reducing the effective distance to which they would otherwise be dispersed. Darkness may contribute to the longevity of certain organisms and enhance the distance to which organisms are dispersed. Situations may arise, however, in which winds, rains, or other factors favorable to dispersing organisms are not coincident with the time in which such organisms are dispersing.

TEMPERATURE

Temperature is essential for the dispersion of organisms. Fatally high temperatures occur at 45°-50° C. and destroy individuals. Some protected spores and seeds survive temperatures as high as 100° C. for a short time. *Autoclaving* is the practice of using 121° C. for 20 minutes or more to kill many infesting organisms. Most insects perish at 50° C. after a few minutes of exposure. Critical temperature ranges affect the dispersal activities of many species, although wide variations are tolerated.

Temperature effects may depend somewhat on whether the disseminules disperse actively or passivly. Passive disperser organisms are apparently affected less by high or low temperatures than are active disperser organisms, except indirectly through organisms that are agents of dispersion and that are affected by temperatures.

The discharge distances of the eight-spored projectile of *Sordaria fimicola* were longer at higher than at lower temperatures. Average distance of the discharges were 5.23 cm. at 7-10° C., and 6.30 cm. at 21-40° C., according to Ingold and Hadland (1958).

More individual insects and more species were taken with rising temperatures in the trapping reported by Freeman (1945). The total numbers of insects collected over the temperature ranges are summarized in table 121.

Table 121

Insects collected at temperatures and elevations
(data from Freeman, 1945)

Temperature (° F.)		42-52	52-64	65-72	73-83
Height collected (ft.)	177	18	46	52	71
	277	10	25	35	36
Totals		28	71	87	107

Successively more insects were taken with successively higher temperatures from 43° to 83° F.

Temperature affected the maximum distance of dispersal, in feet, of the citrus red mite, according to Tashiro (1966). Maximum distances of dispersion in warm (June-July) and cool (March-April) periods were taken from the author's Fig. 2 and are given in table 122.

Table 122

Dispersion of citrus red mite during warm and cool periods
(data from Tashiro, 1966)

Temp. during per.	Days after release										
	0	1	2	3	4	5	7	8	10	14	18
Warm	1.5	1.5	2.0	—	4.0	—	—	4.0	5.0	—	8.0
Cool	0.1	1.0	—	2.0	—	2.0	2.0	—	—	3.0	3.0

Mites dispersed at least twice the distance in warm as in cool weather.

There is less aggregration of the white pine conebeetle, *Conophthurus coniperda* (Schwarz), at lower temperatures, according to Henson (1961) (table 123).

Table 123

Temperature effects on index of aggregation
of white pine conebeetle
(data from Henson, 1961)

Temperature, ° C.	Index of aggregation*	Error
15	0.620	0.105
25	0.447	0.056
35	0.401	0.042

*An index of complete aggregation is 0; that of random association is 0. Greater aggregation of weevils is not unexpected at the higher temperatures, since more biological activity is usual at higher temperatures than at lower temperatures.

Temperature and rate of biological actions may be analogous to chemical actions. The acceleration of chemical reactions has been expressed by van't Hoff's law, in which the velocity of chemical actions is increased with a unit of increase in temperature. It has also been called the "Q_{10}law." As applied to the dispersion of small organisms, it could indicate that for each unit, say 10°, rise in temperature the distance of dispersal would be increased by some factor, say two-fold. Reversely, lowering of the temperature would reduce basic metabolic function and thereby dispersion activities. There is undoubted and generally recognized knowledge that changes in temperature do alter the activities of small organisms. It is a common observation that mosquitoes are not active in biting at temperatures as low as 60° F., for example, but that as the temperature rises to 70° or more, biting becomes active. A discussion was given by Rudolfs (1923) on the alighting of mosquitoes, in which the Q_{10} factors ranged from 1.1 to 1.5 over the temperature ranges of 5° classes from 15° to 30° and when the wind was 4, 4-8, and 8 miles per hour.

A sound philosophy was expressed by Rudolfs (1923):

It is possible also that different influences (humidity, light) may be limiting factors at different temperatures and that therefore the constructed curves are made up of several different parts, of which each is a perfect van't Hoff curve.

Dispersion of *Drosophila pseudoobscura* Frowola flies was found very much affected by temperatures (Dobshansky and Wright, 1943). There was little dispersal below 60° but the distance of movement increased with higher temperatures, almost 28 meters per degree. They observed decreased dispersal with increasing age of flies.

Glick (1939) and Freeman (1945) have shown that ground surface temperatures are related to the abundance of insects at different elevations. Insects were collected at altitudes of 200 feet by day and of 500 feet by night flights (Glick, 1939), by which data taken from Fig. 5 indicate the comparative insect abundances (table 124).

Table 124

Insect abundances collected by day and night at various temperatures
(data from Glick, 1939)

Temperature (° F.)	40	45	50	55	60	65	70	75	80	85	90	95
Collection, day	2	4	6	6	9	12	16	18	17	15	11	7
Collection, night	—	—	—	11	15	14	15	18	16	14	—	—

Most insects were taken at 75° F., regardless of whether the collections were made by daytime or nighttime flights.

Population densities of insects increased with temperatures over the 40° range, 43-83° F. (Freeman, 1945). This is shown in table 125.

Table 125
Insects collected by temperatures at elevations
(data from Freeman, 1945)

		Altitude						
	177 ft.				277 ft.			
Temp. (°F.)	43-52	52-64	64-71	73-83	43-52	52-64	64-71	73-83
Hemiptera	4	22	22	26	4	13	17.5	15
Hymenoptera	0.4	9	8	11	0.3	5	4	6.5
Total insects	18	46	52	71	10	25	35	36

A greater response to temperatures is seen at the lower altitude, in which comparatively more insects were taken at the higher temperatures in the lower strata. A difference of 30 degrees (43-52 to 73-83) increased collections 4.0 to 27.5-fold more at 177 feet and 2.6 to 21.6-fold more at 277 feet. Aerial populations are more responsive to temperatures at the lower elevations. A summary of the above data, altitudes combined, is given in table 126.

Table 126
Insects collected at temperatures, altitudes combined
(data from Freeman, 1945)

Temperature (°F.)	43-52	52-64	64-71	73-83
Hemiptera	8.0	35.0	39.5	41.0
Hymenoptera	0.7	14.0	12.0	17.5
Total insects	28.0	71.0	87.0	107.0

Insect collections made by a rotating net (Lewis, 1950) were related to temperature maxima, as given in table 127.

Table 127
Insects collected at maximum temperatures
(data from Lewis, 1950)

Temperature (°F.)	75	78-79	82	85	88	93	100
Insects collected	128	102	171	120	221	176	217

These data were related to the passage of frontal systems, however, and these in turn were related to barometric pressure, which is discussed on p. 137. It is noted that most insects were taken at 88° F., nearly as many at 100° F., and that fewer individuals were caught at temperatures of 85° to 75° F.

Existing temperatures are important factors in the biting activities (a definite dispersion activity) of mosquitoes. Numbers of mosquitoes, *Aedes sollicitans* (Wlk), with a few *A. cantator,* alighting were related to temperatures, given by Rudolfs (1923) in table 128.

Table 128
Incidence of mosquito alighting at temperatures
(data from Rudolfs, 1923)

Temperature (°F.)	60	65	70	75	80	85	90
Mosquitoes alighting, wind 0-4 mph.	4	19	41	41	44	68	75

Although temperatures were favorable for mosquitoes to alight, winds above 9 mph reduced the alighting numbers.

The influence of temperature and relative humidity on the flight performance of tethered virgin female *Aedes aegypti* (L.) was investigated by Rowley and Graham (1968). Distances flown by *A. aegypti* at different temperatures were given. Most distances near 13,000, 14,-000 and 15,000 were obtained with little differences associated with 30%, 50%, and 90% relative humidity. Temperatures of 15°, 18° and 27° reduced distances somewhat. Only short distances were flown at 10° and 35°. Speed and distances appear to have been altered by temperatures.

Migrations of the silver-Y moth, *Plusia gamma* L., were found by Fisher (1938) to be related to a steady or rising temperature. Through combining directions (no easterly flight was recorded) this is shown in table 129.

Table 129
Silver-Y moth flights related to change of temperatures
(data from Fisher, 1938)

Thermometer	*Total number of flights observed*
Rising	21
Falling	7
Little change	20

The rate of flight of the alfalfa butterfly, *Colias philodice eurytheme* Boisduval, was determined by Leigh and Smith (1959). They found the rate of flight after the early morning hours until sustained flight cessation in the afternoon was nearly constant but that in the other hours flights were erratic. Rates of flight as related to temperatures by pooled data are given in table 130.

Table 130
Flight speeds of the alfalfa butterfly at temperatures
(data from Leigh and Smith, 1959)

Air temperature (°C.)	19-23	24-26	27-29
Rate of flight (ft./min.)	304	320	338

Rates increased with increased temperatures. These temperatures coincided somewhat with the middle daylight hours; although there is no known evidence of it, this factor may have had an influence on rate of flight.

WIND AND AIR CURRENTS

Air currents are present everywhere and affect all organisms. These air movements affect insect ecology very extensively, according to Wellington (1954). Movements of air are notable for their variations. These variations range from calm periods over long intervals (hours) to very infrequent wind speeds in excess of 200 miles per hour in extremely localized areas for very short time intervals (minutes). Such variations are undoubtedly very important and fundamentally essential to the dispersion of organisms. Two characteristics of wind—speed and direction—are the important ones.

A conical shaped pattern of distribution was discussed by Wilson and Baker (1946a) as ideal for the dispersion of fungus spores that originated from a point source. Such an ideal would occur only if the spores dispersed in such relationships that spore densities were equal at all points in vertical planes and if the pattern were circular. A line or areal source of spores would give a large number or a continuum of overlapping cones downwind from the spore origin. Although such idealistic models cannot occur in nature, owing to many factors that cause distortion and change of air movements, they afford patterns by which to study organismal dispersion. Conceivably others—Stepanov (1935) and Gregory (1950), working with fungus spores, and Johnson (1950), working with aphids—recognized the tendency for cone-shaped distribution patterns of organisms to occur during dispersion.

Air movments are responses to temperature or pressure differences and tend to equalize these differences, although for organisms equalization is never quite complete. Turbulent movements bring together different air masses, which mix and dilute groups of dispersing organisms Continued turbulence disperses organisms widely and distantly, giving rise to constantly diminishing populations (Gregory, 1945). During periods of calmness followed by turbulence or sudden gusts, impacts engage organisms in transport to bear them upward and away from the original site more than do continuous or high velocity winds. This is done in a manner illustrated by Öort (1940) for spores of loose smut of wheat, *Ustilago tritici* Pers. (fig. 31).

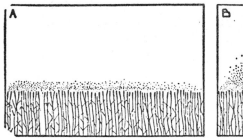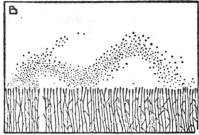

Fig. 31. Dispersal of smut spores by winds. A—Strong wind. B—Weak wind (schematic diagram, fig. 2 reproduced from Öort).

Low velocity winds apparently spread spores widely rather than distantly; low velocity winds appear to be more closely associated with turbulences and may acount for more widespread spore distribution than do high velocity winds. Media favorable for the growth and reproduction of organisms are likely to occur near the source. Low velocity winds with turbulences and cross currents may be more favorable to the dispersal of organisms than are continuous high velocity winds.

Deposition of *Lycopodium clavatum* spores was studied by Gregory (1951) in wind tunnels in which turbulent and streamlined winds of five different speeds moved spores to impinge on coated glass cylinders perpendicular to the tunnel axis, or the windward direction. Results were obtained as spores deposited on glass cylinders 0.53 cm. in diameter and are shown in figure 32 as profiles of crosswind distributions. Some general observations indicate that turbulent winds tend to provide more uniform deposits, perpendicular to the axis above and below the axis of the tunnel, than streamlined winds. Nearly 3,000 spores were found at the axis, while about 1,300 to 1,500 were taken at +4.25 cm. above or below the axis by a 9.4 m./sec. turbulent wind. Over 7,000 spores were deposited at —4.25 cm. below the axis compared with less than 1,000 at + 4.25 cm. above the axis by a 9.0 m./sec. streamlined wind. Spore depositions in the streamlined wind tunnel are indicated by lines with many acute angles with the tunnel axis. Those deposits in the turbulent stream are more perpendicular to the axis than are those of the streamlined winds. This apparently has a significant bearing on dispersion by winds. Organisms are dispersed more by slight turbulences than by straight winds (Wellington, 1945). Energized irregular turbulent thrusts by eddy currents may move more organisms farther than strong straight winds.

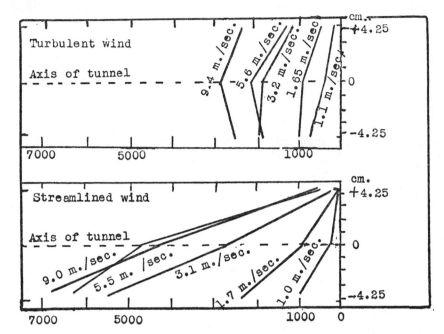

Fig. 32. Vertical crosswind *Lycopodium* spore distributions from turbulent wind currents (top) and streamlined winds (bottom). Ordinates indicate distances above and below the axes of tunnels and indicate spore deposition on 0.53 cm. diameter glass cylinders, per cm². Abscissae indicate number of spores per 10^6 liberated at 1.4 m. distance, for each of five wind speeds indicated (adapted from Gregory, 1951).

Turbulent forces are dependent on the rate of increase of velocity with height. Such forces of wind against the organisms possess "drag velocity" and are influenced by surface roughness. Surface roughness is dependent on the size, character, arrangement, situation, and position of objects on the earth's surface. Open areas composed of comparatively small objects, such as low ground coverings or buildings, give rise to turbulences. Turbulences, reasoned Durham (1951), tend to hold pollen grains in suspension until they are caught by ascending currents and carried upward.

Strong air currents move downward and return organisms to the earth more rapidly than do calm periods, during which only gravity

exerts sedimentation processes. Constant movements, unequal displacements, and mixing of air with equally variable movements permit displacements of organisms carried by air currents.

Tangential currents may be more fully understood by the construction of a cone with its apex at some height, say 7.5 feet, above ground
level and its base a perpendicular plane 15 feet in diameter at a distance
of, say 30 feet from the apex. Then pass a vertical plane through the
apex and the center of the base parallel to the predominant wind direction. A horizontal plane may pass through the apex and the center of
the base. The apex of the cone may be understood as a source of the
organisms and the base as a target area. Passive disperser organisms released at the source move toward and strike the target, giving characteristic patterns. One such pattern ("smoke" clouds) is given in figure 33
from Wilson and Baker (1946). Particles struck the target, giving an
ovoid pattern which is apparently one of the most significant features of
tangential dispersion. More particles (43) struck below the horizontal
line, fewer (35) above the line. Slightly more particles (40) struck the
right side of the target, slightly fewer (38) struck the left side of the
target. Particle spread is greater horizontally and lesser vertically. Gravity

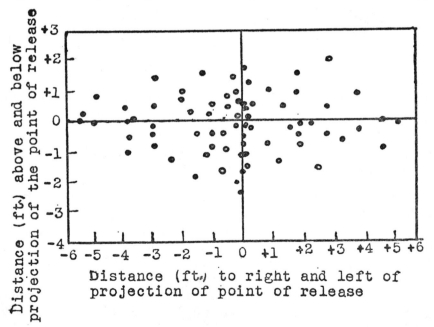

Fig. 33.   Ovoid pattern dispersion with a wind velocity of 6.4 miles per
hour, 4.5 feet above ground and 16 feet from the origin (fig. 3 reproduced
from Wilson and Baker).

(see figs. 14, 17) may account, in part, for more particles below the center line.

Effects of tangential currents were measured in angles as related to the predominant wind direction (Stepanov, Test II, 1935). Spore counts of *Tilletia tritici* (Bjerk.) Wint. were combined from the 5, 10, 15, and 20 meter distances, from wind speeds of 0.5 to 4.0 m./sec. for each angle degree given in table 131.

Table 131
**Spore deposits at angle degrees from
the predominant wind direction
(data from Stepanov, 1935)**

Angle (degrees)	Total no. spores per cover glass
—45	4
—30	131
—15	103
0	424
+15	771
+30	1150
+45	1218
+60	299

Close relationships are lacking for the predominant wind direction, 0 degree, and the angle with greatest abundances of spores. In another test (Test IV), combining data from the distance range of 94 meters to and including 1,034 meters, results of spores collected on 18 x 18 glass slides are given in table 132.

Table 132
**Spore deposits at predominant wind direction
and at angles
(data from Stepanov, 1935)**

Angle degrees	Total no. spores
+15	146.2
0	231.3
—15	125.8

These data indicate a close relationship between tangential wind direction and the number of spores collected. Since the above observations were made under natural or field conditions, some rather wide variations are not unexpected.

Distributions of *Lycopodium* sp. spores were greatest at the level of release and decreased upward and downward to the periphery of the experimental conical model used for the collections. Spore collections were made above and below the level of release at seven different wind velocities (Wilson and Baker, 1946a). Combinations of the data taken at 5, 10, and 15 feet for each wind speed are used to show the results. These are given as spores per 67 mm.^{2} slide in table 133.

Table 133

Lycopodium spore deposits during wind speeds at distances
above and below the level of spore release
(data from Wilson and Baker, 1946a)

Feet above and below level of spore release	Wind speed (mph.)							Total
	3.8	5.9	6.4	8.1	8.5	13.4	16.0	
+6	21	0	0	1	2	0	3	27
+5	27	9	0	1	5	11	11	64
+4	23	25	3	7	10	18	20	106
+3	44	38	11	37	52	53	99	334
+2	44	69	40	104	119	117	262	754
+1	151	387	169	324	414	332	1042	2819
0	269	1242	406	988	1274	1020	2070	7296
−1	169	260	102	566	446	480	1128	3151
−2	103	53	21	171	116	123	278	865
−3	45	20	4	51	26	25	98	309
−4	34	2	2	9	4	7	10	68
−5	16	5	0	1	5	1	0	28
−6	7	0	0	0	0	0	1	8
Totals	980	2110	758	2259	2473	2187	5022	15789

Many spores were taken at the spore release point level, 0 feet, and are indicated by the data at each wind speed. Tangential effcts are indicated as having more extensive and wider spread of spores with winds of lesser than of higher speeds. Such effects are congruent with the observations of Öort (1940). The total number of spores caught above the 0 level was 4,104, and the total below 0 was 4,429. The difference, 325, may be partially accounted for by gravity.

Although there are some directional effects, they are not as extensive as might be expected. In general, spores, seeds, pollen, and other passive disperser organisms are dispersed omnidirectionally as measured by end results over long periods. Organisms, in most instances, may disperse unidirectionally for a short period. Owing to tangential currents and to changing air movements over long periods, however, omnidirectional dispersal achievements are more frequently observed than are those in only one direction. This must occur, apparently, for the essential economy of all species. Such dispersal would move organisms into areas that

are favorable and unfavorable to them where many would live and reproduce, although many would be lost. Colonies would be provided for
nearby and somewhat distant areas with individuals that might find
suitable media in which to live and to reproduce. Dispersal would be
made into fringe areas where species might live and reproduce during
favorable periods and perish during unfavorable periods.

Eddy or turbulent currents are comparatively large bodies, not small
quantities, according to Wilson and Baker (1946a). They maintain unity
for some time and vary more in horizontal than in vertical movements.
Comparisons of vertical and horizontal dispersion of *Lycopodium* sp.
spores were made above and below the horizontal plane containing the
release point and to the right and left of the vertical plane containing the
release point and parallel to the wind direction. These comparisons were
measured by the standard deviations of spore distributions at distances
from the release points and at different wind speeds given in table 134.

Table 134

**Standard deviations horizontally or vertically
to spore release points at wind speeds
(data from Wilson and Baker, 1946a)**

| Wind speed (mph.) | Direction | Standard deviations at distances from release point | |
		5 ft.	15 ft.
3.8	Vertical	0.62	1.59
	Horizontal	1.8	2.48
6.1	Vertical	0.61	1.81
	Horizontal	1.47	2.39
7.2	Vertical	0.59	1.55
	Horizontal	0.94	2.32
10.3	Vertical	0.55	1.53
	Horizontal	1.08	2.81
Averages	Vertical	0.59	1.62
	Horizontal	1.14	2.50

Further work by Wilson and Baker (1946a) showed that "smoke" clouds
spread more widely horizontally than vertically, furthermore they tended
to describe patterns that were oval and flattened horizontally (fig. 33).

Active disperser species are affected by winds in two ways, although
both are results of the forces acting against the organisms. One way is
that in which winds, at least milder ones, supply energy and aid in the
successful dispersion of many species. Such species accept the transportation by slower air movements in order to feed, mate, and reproduce,
and they cannot be blown far from their course. In the second way, which

is restrictive, air movement accelerates and progresses beyond the speed at which the individuals are no longer able to maintain their position. At such speeds most insects come to rest in protected places and remain until the wind velocity is reduced and they can again control their movements.

Active disperser organisms frequently possess passive disperser phases. In these instances the passive disperser phases disperse similarly to organisms that are wholly passive.

Wind tunnel. Wind tunnel studies are worthwhile and are needed. They suggest how field activities may occur and they may explain the behavior of some observed dispersion phenomena. Wind tunnels, however, doubtless alter light, humidity, temperature, turbulences, and other factors, so that caution is required in transferring results obtained to field activities.

Very thin strata of air that are very near and in contact with tunnel walls remain at rest, resisting movement. This precludes uniform velocities within the tunnel. With a wind speed of 10 m./sec at the tunnel axis the velocity was 10-12% higher at about 1½ inches from the wall, and at distances of 2 inches there were variations of 5-6% (Gregory, 1951). Hence, it is seen that wind speeds in tunnels are variable. Variability was increased if the speeds outside the building were in excess of 3 m./sec. (Gregory, 1951).

Summary statements covering some of the more important findings and conclusions from Gregory (1951) are given as follows:

1. In movement and deposition, individual spores act independently of one another.
2. Dilution by turbulences is independent of wind speed.
3. Streamlined currents deposited more spores at the axes of wind tunnels than did turbulent currents.
4. Turbulent currents deposited more spores on small (0.02 cm.) adhesive coated cylinders than on those of larger (to 2.00 cm.) sizes in all wind speeds tested (1.1-9.7 m./sec.).
5. Most spore deposits on the surfaces of cylinders impacted on the midline of the cylinder facing the wind, became fewer toward the sides, and became zero in the areas on the edges tangential to the wind direction.
6. Percentage efficiencies of commonly used spore traps were greatest with a 0.53 cm. diameter cylinder, ranked secondly with a 1.4 cm. cylinder, and were least with vertical slides of 1 x 3 inches

Apparently there have been no wind tunnel studies with pollen, seeds, nor insects.

Wind speeds. Different wind speeds may have different effects on dispersion of the same species or the same effect on different species.

It is difficult to measure effects of winds under field or uncontrolled conditions, since speeds are very variable and it is infrequent that sustained air currents are suitable for conducting experiments. Such variability tends to discourage measurements for test purposes and reduces the accuracy of those that are attempted. Observations have been recorded, however, under natural conditions in which it was practical to report predominant wind speeds and which contribute to knowledge. Wind tunnel, i. e., laboratory or controlled, studies have been made and give considerable information on effects pertaining to some species of organisms. The dispersion of salt water droplets has been studied and provides some information on wind speeds as they affect movement of this lifeless material.

Vertical and horizontal dispersion of salt water is recognized. Plants growing near shore are often injured and killed by such salt water, especially following winds of unusually long duration and high velocities. Doubling of wind speed may result in deposits of more than twice as much salt. This is seen in figure 34 with regression curves computed by data given by Boyce (1954). These curves show more salt deposits at the nearer distances, as would be expected, and they show something of the amounts of increase with increased wind speeds.

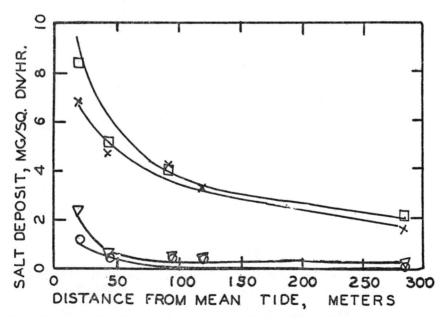

Fig. 34. Incidences of salt deposits at distances from the seacoast. Curves from lowest to highest positions represent different wind speeds, 2.5, 5.5, 8.0 and 11.0 m., respectively (data from Boyce).

Effects of prevailing winds in the incidence of BMYV were shown by Hampton (1967, fig. 35) in forming incidence patterns. Although the initial incidence percentage was nearly the same against the prevailing wind as with it, a distinctly more rapid slope was found on the side against the wind. A slightly more gentle slope of the curve downwind and more distant (over twofold) distribution is shown (fig. 35). Incidence against the prevailing wind appears to reach zero near 350 feet and with the wind to exceed 1,000 or 1,200 feet from the clover fields.

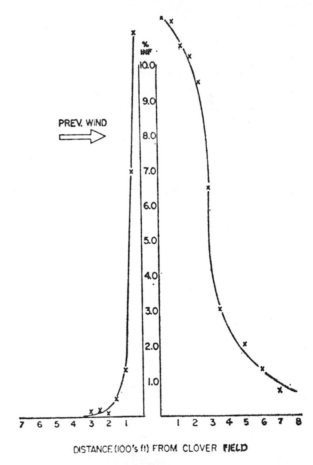

PREV. WIND

DISTANCE (100's ft) FROM CLOVER FIELD

Fig. 35.   Incidence patterns of BMYV on distance from red clover fields as influenced by prevailing winds (fig. 3 from Hampton, 1967).

Density of *Lycopodium* sp. spore dispersal at different wind velocities at three right plane sections (at 5, 10, and 15 feet from the apex) of the experimental cone model were given by Wilson and Baker (1946a). Observed data are given in table 135-136.

Table 135-136
Lycopodium spore deposits at feet from the release point
and at different wind speeds
(data from Wilson and Baker, 1946a)

Wind velocity (mph.)	Spores per 67 mm² at feet from cone apex		
	5	10	15
2.2- 3.8	360	102	57
3.8- 5.1	533	155	64
5.5-11.3	936	933	867
12.1-16.0	867	300	326

An order of increase in the number of spores caught with wind velocity increase is seen to about 12 mph. at 12.1 to 16.0 mph. Decreased numbers of spores were found at 5, 10, and 15 feet from the cone apex.

Spores of *Tricothecium roseum* Link were colected at 31% R. H. at different wind speeds by Zoberi (1961). Data on these collections are given in table 137.

Table 137
Spores collected at different wind speeds
(data Zoberi, 1961)

Speed (m. sec.)	1.7	2.5	3.3	5.0	6.7	10.0
Spores per tube (no.)	132	399	1,234	2,358	5,879	12,375

Increasingly more spores are taken with increasingly higher wind speeds.

Decreasing numbers of *Lycopodium* sp. spores were found with decreasing wind speeds. More spores were at the axis as dispersed by streamlined than those dispersed by turbulent winds, (Gregory, 1951). This is attributed to the displacement of the spores by gravity.

More *Lycopodium* sp. spores impacted on cylinders of small (0.018 cm.) diameter than on larger (2,000 dm.) adhesive coated cylinders at each of five wind speeds. Data from each cylinder size at different wind speeds were summarized by wind velocities as mean number of spores per cm.² on the axis of wind tunnel, from Gregory (1951), and are given in table 138.

Table 138
Impaction of *Lycopodium* spores by wind speeds
(data from Gregory, 1951)

Wind speed (m. sec.)	1.1	1.75	3.3	5.6	9.7
Spores per cm.²	992	1281	1931	2405	2573

Cylinders of smaller diameters are more effective in trapping spores than cylinders of larger diameters, and more spores are impacted by successively higher wind velocities.

Further data on the dispersal of *Lycopodium* sp. spores as affected by wind speeds are given above under *Tangential Currents*. A recapitulation of those data, spores caught at each wind speed, is given in table 139.

Table 139
Lycopodium spores taken at wind speeds
(data from Gregory, 1951)

Wind speed (mph.)	3.8	5.9	6.4	8.1	8.5	13.4	16.0
Spores caught (no.)	991	2112	758	2259	2469	2187	5020

An order of increase is shown for wind speeds from 3.8 to 16.0 mph., with exceptions at 6.4 and 13.4 mph. These are variants not uncommonly observed in biological research.

Measurements of spore dispersion may be accomplished by computing standard deviations. Such measurements were reported by Wilson and Baker (1946a). They stated that "Effects of wind velocity must be looked for accordingly, in the data on spore dispersion, not in the data on comparative densities." (Note that dispersion as used here refers to the mathematical meaning of scatter, spread, or variation—not to the "dissemination" or "exodus."). Mean standard deviations of vertical distributions for the different wind velocities from Wilson and Baker (1946a) are in table 140.

Table 140
Standard deviations of spore deposits at wind speeds
(data from Wilson and Baker, 1946a)

Wind velocity (mph.)	1-3	3-5	5-7	7-9	9-11	13-15	15-17	17-19
Mean standard deviation	1.57	1.13	0.58	0.64	0.45	0.62	0.59	0.45

Variations were wider in lower than in higher velocities. At medium to high velocities the variations were small and nearly equal. Scatter or spread of spores appears more important at wind speeds of less than five miles per hour than at higher speeds. Such scatter exhibits more tangential effects through the spread of spores than do winds of higher velocities.

Wind speeds doubtless influence pollen deposition, although few comparative data taken under field conditions measure wind speeds as they affect pollen dispersion. Pollen may or may not respond to wind speeds as fungus spores do. Evidence available indicates that winds having higher speeds disperse more pollen longer distances than do those of lower wind speeds. Considerably more pollen was deposited by the higher speed winds at the 100 meter distance than by weaker winds (fig. 36, from Jensen and Bøgh, 1941). Clearer evidence of effects

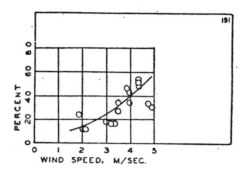

Fig. 36. Pollen dispersion as related to wind speed (fig. 3 from Jensen and Bøgh).

of strong winds, moderate winds, and calm periods on the dispersion of pollen from swedes, *Brassica* sp., by Jensen and Bøgh (1941, table 13) is given in table 141.

Table 141

Pollen deposits by strong, moderate and calm wind periods
(data from Jensen and Bogh, 1941)

Distance from pollen source (m.)	Percentage of pollen		
	Strong wind	Moderate wind	Calm
0	100	100	100
5	145	39	—
10	72	31	49
20	22	13	6
40	30	10	0
60	66	6	0
100	66	2	0

Effects of wind speeds on a chironomid, *Spaniotoma (Orthocladius) perennis* Mg., an insect having a comparatively weak flight, were determined during tests comparing the performances of two traps (table 142, Johnson, 1950).

Table 142
Chironomids collected at different wind speeds and hours
(data from Johnson, 1950)

Wind speed (mph)	1.4-2.1	2.1-2.9	2.9-3.7	3.7-4.6	4.6-5.5	5.5-6.5	6.5-7.4	7.4-8.4	8.4-12.0
10 A.M. suction trap	301.5	124.9	73.4	40.3	32.2	11.1	2.0	1.9	4.3
Midnight sticky trap	48.6	35.1	16.7	6.8	8.3	3.6	0.4	1.4	2.6
10 P.M. suction trap	76.6	37.1	23.0	18.4	11.7	4.0	0	3.6	4.6
Midday sticky trap	12.8	17.5	7.1	7.7	5.6	6.2	0	1.7	1.5

Differences are shown with different wind speeds, regardless of period of collection or of trap. Fewer insects were taken at the higher wind speeds and more at the lower, 6.5-7.4 mph., wind speed classes.

Wind velocity affects the density of spiders, Araneida, to 300 feet altitude, according to Freeman (1946). Relationships shown from 10, 177, and 277 feet altitude collections are indicated in table 143.

Table 143
Spiders collected at different wind speeds
(data from Freeman, 1946)

Wind speed (mph)	8	11	18	23	25
Spiders collected (no.)	5.9	1.9	0.6	0.6	0.5

Most spiders were taken at the lowest of the wind speeds over the range, 8 to 35 mph. This is in agreement with the work from Glick (1939), who reported that most spiders were taken at 3 to 4 mph., although many specimens were taken at 7 to 8 mph.

Davies (1936) found that green peach aphids, *Myzus persicae* (Sulz), did not initiate flight when wind speeds were as high as 3.75 mph. The aphids averaged 154.8 flights per minute during calm periods. Wind below 5 mph. was favorable for aphid flight, according to Thomas and Vevai (1940).

Wind velocity was related to the number of insects collected (all species), as measured at different elevations by Glick (1939). Surface velocities appear to have affected collections at 200 feet more than at higher elevations. A peak of abundance was found at 200 feet elevation with wind speeds of 5-6 mph. At 1,000 and 2,000 feet, peaks of abundance were at 7-8 mph. At 3,000 and 5,000 feet much flatter curves were shown by Glick (1939), indicating that surface winds were not very influential in affecting the numbers of insects collected. High velocities (45 mph.) at altitudes of 6,000 to 16,000 feet were believed of much importance by Glick (1939), since they carry insects upward and to great distances. There are no known data, however, that show the frequency nor extent of such occurrences.

There is much variation among orders of insects as to flight during wind speeds. Populations of Hymenoptera, for example, increased in abundance to a peak at speeds of 21 mph., although populations of Coleoptera declined at speeds in excess of 6 mph., rapidly to 10 mph., then slowly at higher speeds (Freeman, 1945). Populations, total insects collected, were listed at different ranges of wind velocities by Freeman (1945), as shown in table 144.

Table 144
Insects collected at wind speeds and at two altitudes
(data from Freeman, 1945)

Altitude (feet)	Wind speed (mph)				
	6-9	10-12	14-21	22-25	35
177	46	68	51	20	6
277	31	32	30	14	2

Most insects were collected at the lower altitude and at wind speeds of 10-12 mph., regardless of altitude.

Wind direction effects on a pentatomid, *Aelis rostrata*, were studied by Brown (1965). Data presented are given in table 145.

Table 145
Pentatomid dispersal against, with, and across wind directions
at initiation and termination of dispersal
(data from Brown, 1965)

Flight	Direction in relation to the wind	Against	With	Across
Initiation:	all records (no.)	199	450	435
	(%)	18.4	41.5	40.1
	Wind 6 ft./sec.			
	plus (no.)	25	135	61
	(%)	11.3	61.1	27.6
Termination:	all records (no.)	129	653	309
	(%)	21.8	59.9	28.3
	Wind 6 ft./sec.			
	plus (no.)	11	157	60
	(%)	5.0	67.9	27.1

Most insects initiated and terminated dispersion with the wind; fewest dispersed against the wind. Winds over 6 mph. increased the percentage over the "all records" groups. Winds of 6 mph. plus increased the percentages "with" and decreased the "against" and "across" direction of air movements.

In migration studies of the silver-Y moth, *Plusia gamma* L., Fisher (1938) reported on flights with, against, and across winds at different wind speeds, as given in table 146.

Table 146

Silver-Y moth migrations related
to wind currents and speeds
(data from Fisher, 1938)

Flights						
	1-3	4-7	8-12	13-18	19-24	*Total*
With wind	4	3	2½	—	—	9½
Against wind	4	4	3½	½	—	12
Across wind	5½	6½	2	1½	2½	18

The most flights were by cross winds at most speeds, the least were with the winds. Most flights occurred at low, 1-3 and 4-7 mph. speeds.

A feature of the work by Bailey and Baerg (1967) was that in relation to crosswind, downwind, and upwind, with data given in table 147.

Table 147

Mosquito collections with relation to winds
(data from Bailey and Baerg, 1967)

Wind	Up-	Down-	NE cross-	NW cross-
Mosquitoes per collection	105	131	92	93
Marked mosquitoes recaptured (no.)	1	4	1	1

Collections downwind were slightly more abundant than those in the other quadrants. Collections upwind were, however, slightly more than those crosswind. A statistical analysis of the data given showed that the interaction of collections by months, September through March, a seven-month period, and wind direction varied no more than might be expected by chance occurrence.

In studies to determine (1) the maximum wind velocity against which the mosquito, *Culex tarsalis* Coq., could fly and (2) the maximum distance it could travel in one evening with the aid of the wind, Bailey *et al.* (1965) found that mosquitoes dispersed almost equally over hills and valleys without evidence of retardation or barrier effects. Certain other observations were given without benefit of regression comparisons as follows:

Wind speeds of 3 or 4 mph aided in moving mosquitoes downwind but 6 mph tended to inhibit dispersal.

A recorded movement of 5 miles downwind on the night of release and of 15.75 miles two nights later was found.

Speed of flight was calculated at 4.75 mph over short distances.

Retardation effects of artificial windbreaks in the distribution of insects to the leeward of sheltered zones were reported by Lewis and Stephenson (1966), although effects of shelters have long been recognized by entomological workers on localized abundances of insects. Windbreaks of varying permeability to air currents were erected to compare 0% (solid), 25%, 45%, and 70% wind passages. Suction traps collected insects about the windbreaks only when winds were blowing from directions to encounter the barriers from 90° directions or less. Mean percentage increases in the number of insects caught at nearby exposed sites are tabulated by families in table 148.

Table 148
**Insects of different families aided or retarded,
based on permeability of fences
(data from Lewis and Stephenson, 1966)**

Insect family	Permeability of fences (%)			
	70	45	25	0
Aphidae	-7	13	30	32
Cecidomyiidae	25	43	225	260
Mycetophilidae	15	105	348	250
Mymaridae	22	60	387	479
Phoridae	23	27	90	64
Psychodidae	80	24	16	870
Staphylinidae	857	761	632	571
Thysanoptera (order)	126	38	90	178
Parasites (order of Hymenoptera)	15	4	27	97

More insects accumulated nearer to solid windbreaks, or barriers, than to those areas permitting wind passage. An inverse relationship was indicated, however, for species of the Staphylinidae. Less degree of response to permeability is indicated by species of Phoridae (from 23 to 64, nearly threefold.) than to Mymaridae (from 22 to 479, almost twenty-twofold) in congregations near the barrier.

Relative densities of insects were shown by Lewis (1967) fig. 7, reproduced as fig. 37, to windward and leeward of a 45% permeable fence. Lath fences three to eight feet in height were constructed and set perpendicular to the wind flow to study patterns of insect dispersion over or through the fence. Most insects accumulated at a distance two to three times the height and leeward of each fence.

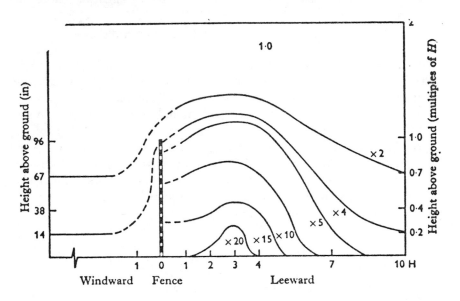

Fig. 37.   Cross-sectional profile showing the relative densities of insects windward and leeward of a 45 percent permeable fence (data from Lewis, 1967).

Barometric pressure. Small organisms are affected apparently not at all or but little by the usual barometric pressures. Absence of reference to the effects of barometric pressure on the dispersal of plant pathogens is observed in two recent reviews of plant pathology, Horsfall and Dimond (1960) and Stakman and Harrar (1957). No reference was observed concerning the effects of barometric pressure and distance on the dispersal of pollen or seed. There are references to insect behavior, however, and some will be given.

Barometric pressure variations are associated with other meteorological factors. Lower temperatures, for example, are frequently present with high barometric pressures, and higher temperatures with lower barometric pressures. Accelerated rates of evaporation are usually present with lower pressures and higher temperatures. Higher evaporation rates increase activity on the part of many organisms. Higher temperatures also accelerate activities and hasten maturity of organisms. Precipitation is frequently associated with lower barometric pressures and may favor the successful growth and development of dispersed organisms. The dispersal behavior of organisms may be only coincidental with barometric pressures.

Barometric pressures are often related to other atmospheric phenomena. Since unstable weather, storm centers, or *air mass* seems to include the various factors, some discussion is given to air masses and their movement. Factors involved in air mass movements include wind speeds, cloudiness, relative humidity, temperatures, precipitation, and barometric pressure, the measurement of which is needed for more complete understanding.

Populations of organisms, especially of insects, according to Wellington (1954), may be affected considerably by air mass movements. Wind, rain, lowering temperature, rising barometric pressures, and other factors attend air mass movements and effect organisms. Many individual organisms will be destroyed by weather factors but few will disperse or be dispersed by the air mass movements to develop and reproduce.

A review of physical factors affecting insects (Uvarov, 1931) indicated that confinement in low pressure chambers for short periods was harmless to many kinds of insects. Insects were able to tolerate enormous fluctuations in pressure, such as never occur under natural conditions. The discussion of vertical distribution of insects and spiders by Glick (1939) in relationship to barometric pressures indicates how several factors may be coincident with, be produced by, or produce changes in barometric pressures.

Insects were taken most abundantly in the daytime when the barometer was at 29.85 inches. Insects declined in numbers as the pressure increased to 30.35 inches, as is shown in figure 38 (Fig. 10 from Glick). Spiders, Araneida, were least abundant at 29.85 inches and increased in abundance with pressure increase. Spider collections, according to Glick (1939), may have been related, however, to a change in barometric pressure. Specimens of the orders Coleoptra, Hymenoptera, and Diptera were collected in the closest relationship to barometric pressure.

Rising, falling, or changeless barometer readings were studied by Fisher (1938) as related to migrations of the silver-Y moth *Plusia gamma* L. More flights were recorded as having originated during periods of rising pressure than during periods of falling or constant pressure. This is shown by data (combing data from all directions) in table 149.

Table 149

Silver-Y moth flights as related to barometer changes
(data from Fisher, 1938)

Barometer	Total number of flights
Rising	33
Falling	12
Little change	6

Rising temperature and barometer are thus shown as favorable conditions in which flight originations occur more frequently.

Vinje and Vinje (1955) say, "Diminution of concentration of biological objects appears to be related to cloud base rather than to any definite altitude." Some relationships between cloudiness and numbers of insects collected were reported by Glick (1939), but variations precluded positive statements.

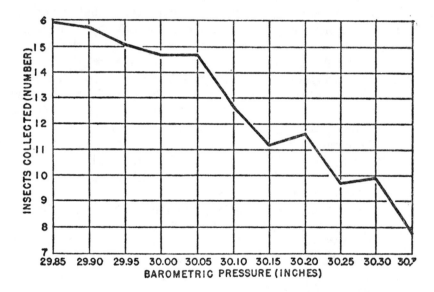

Fig. 38.   Insects collected at different barometric pressures per 10 minutes' flying time in daylight at 200 feet altitude (data from Glick).

Tall trees, thirty to forty feet high, bordering alfalfa fields retarded dispersal of the alfalfa weevil, *Hypera postica* (Gyllenhal), according to Prokopy *et al.* (1967). Such trees apparently reduced retarding effects sharply at the field margin and also at a slight distance, eight feet, from the border, as is shown in table 150.

Table 150
Retardation of alfalfa weevil dispersions by trees
(data from Prokopy *et al.*, 1967)

Feet into woods	Weevils (mean no./sq. ft.) Trees at border	Trees 8 ft. from border
5	2.5	1.2
10	1.2	2.5
20	0.6	1.1
40	0.5	0.5

One-fifth as many weevils were found at 5 feet as at 40 feet, where tall trees bordered the field. At 10-feet compared with 40 feet, where 8 feet distance separated the field border from the trees, there were one-fifth as many weevils. Straw covering on the soil was also shown restrictive of dispersal to estivation sites of the alfalfa weevil, with data as given in table 151.

Table 151
Retardation effects on dispersal of alfalfa weevil at distances
(data from Prokopy *et al.*, 1967)

Feet into woods	Weevils (mean no./sq. ft.) With straw	Without straw
8-12	7.4	2.5
18-22	4.1	1.4
31-35	1.1	0.6

Almost three times as many beetles were found at 8-12 and 18-22 feet into the woods. Straw-covered soil compared with uncovered soil suggested that straw retards dispersing weevils. At 31-35 feet nearly twice as many individuals were taken in straw-covered soil. This suggests that the straw cover had become less of a retarding influence at the longer distance from the field.

Aidants, Hindrances or Barriers

Organisms may disperse in media with or without assistance and with or without difficulties. Aidants may be considered as those materials or factors that assist in dispersion or further the dispersal distance of organisms. Little discussion seems to have been given to this subject. Air movements are positive in moving organisms, but doubtless other aidants are present. Air movement discussions are given under the heading of meteorological factors (chapter 5). More hindrances appear evident than aidants, or perhaps more about hindrances and barriers is recognized. A barrier is a barricade, separator, or isolating mechanism preventing dispersal. It is often considered a total block. A hindrance or impediment is considered a check or retardant of movement.

Doubtless there are aidants, hindrances, and barriers to nearly all dispersal journeys. These may be not recognized, isolated, or definitely measured. Biological control efforts—whether through the use of viruses, bacteria, fungi, insects, sexually sterility (as discussed by Knipling, 1955), or other organisms—depend on dispersal for effectiveness. Aidants, hindrances, or barriers must undoubtedly affect the dispersal of such agents. Better understanding of factors and materials affecting movements of biological control agents could increase the value of released agents.

Dispersal movements of *Lucilia cæsar* L. were reported by MacLeod and Donnelly (1958) as not directed to areas of optimum conditions. In a later article (1962) MacLeod and Donnelly discussed aggregations of blowfly populations and considered that if flies were not restricted by physical barriers they would be expected to aggregate in response to local attraction, although such aidants may not be readily identifiable. Boundaries of aggregation foci were demarked by traps only a few yards apart. It was suggested that microgeographic aggregations might be accounted for by environmental factors or by an "intrinsic character of adult blowfly populations."

Convergence of small organisms appears as much as or even more enigmatical than dispersion. Less discussion and effort has been expended on convergence, however, than on dispersion.

Distance is a very significant hindrance to the dispersion of organisms. Mountain ranges, oceans, and deserts are examples of physical barriers generally recognized by biologists as effective in prohibiting or in restricting movements of small organisms. Shelter is a hindrance to movements, especially of active disperser organisms, as illustrated by Bovbjerg (1952) for movements of a snail, *Campeloma decisum* Say. These were positive upstream movements that were prevented or restricted by stones acting as blocks. Biological, climatological, physiological, and reproductive barriers were recognized by Woodbury (1954). Some aidants, hindrances, and barriers are inconspicuous to man and some may be unknown. Physical barriers influencing movement of the checkerspot butterfly, *Euphydryas edifica* Boisduval, which is extraordinarily sedentary, were absent, according to Ehrlich (1961), from which it was concluded that intrinsic factors or "choice" of individuals limited dispersal.

Media or environments through, around, or among which organisms disperse include fields, forests, orchards, residences, shops, lakes, bags of flour, and many others. All media and the environments through which organisms disperse offer more or less resistance. Successful dispersal movements are related to the total conditions or situations rather than to an individual component or factor.

Air is the most common and important medium through which small organisms disperse. It is also an important aid to dispersion. Components of the air are required in the respiratory processes of organisms and, hence, must be present for maintenance of life. As a rule, more distant dispersion is accomplished by air than by water or soil. Aerial existence is a transient interlude between the initiation and termination of movements of organisms to the earth or to the earth-bound objects.

Water is an agent of dispersion with some species. By its presence or absence, however, water may reduce or increase distances to which organisms may disperse or be dispersed. Water removes and carries organisms from their origin. Rainfall during spore production was suggested by Wilson and Baker (1946a) as a factor reducing the distances to which conidial spores of *Sclerotinia laxa* Ader and Ruh were dispersed by wind. A relationship was reported by Brown (1951) in which species of Corxidae frequenting temporary bodies of water had a higher rate of migration than species frequenting permanent bodies of water.

Although *soil* may be dispersed with organisms, it is seldom an agent of dispersion. Data descriptive of the dispersion patterns of soil-inhabiting organisms are very few. Less is probably known of the dispersion of soil-inhabiting organisms than of those living in the air. Shorter distance ranges are undoubtedly more prevalent with soil-inhabiting species than of air- or water-inhabiting species. Differences in the fluidity of the media may account for the shorter distance ranges. Dispersion through, among, or around soil particles must offer more resistance to and require more energy per unit of distance traversed than does dispersion through air or water.

Airborne organisms were found by Rittenberg (1939) so far from shore that he believed the air was not completely free of organisms anywhere. He found, however, general decreases in populations with distance increases from the origin. Since it is obvious that organisms must decrease in abundance with distance increases, the most outstanding common barrier is distance, according to Elton (1958). Many species are able to achieve worldwide distribution, "either because the ecological barriers that hold in others are not barriers to them, or because, which is partly the same thing, they have exceptionally good powers of dispersal."

Four results of hindrances and barriers are recognized. First, they crowd populations into smaller areas. Second, individuals may reduce crowding by moving into areas outside the hindrances. Third, many organisms attempting to cross the barrier will be lost. Fourth, inbreeding will result in more closely related ontogeny.

Similar types of hindrances may affect different species differently. Bare soil, for example, impeded dispersal of predaceous insects in citrus orchards, according to Carnegie (1957), and this permitted the increase of scale populations in cultivated orchards. Indigenous vegetation, on the other hand, was reported to restrict wind dispersal of scale insect crawlers, thus confining the populations of scale insects. Soil covered but partially by a single row of sunflowers, *Helianthus annuus* L., was reported by Simons (1958) to reduce the veinbanding mosiac virus in pepper, *Capsicum frutescens* L. A 93 percent (or 13/14) reduction was reported in plots protected by sunflower barriers compared with plots without them.

An example of hindrances in kind was given by Jones and Brooks (1950), who discussed corn rows and distances affecting the pollination of corn. An experimental design was used in which a field of a contaminant source corn variety, Yellow Surcropper, measuring 25 x 50 rods, was planted at one end of the field (50 x 100 rods) and blocks of Honey June sweet corn measuring 100 feet square were planted at distances from the Yellow Surcropper (contaminant). Previously harvested small grain fields separated the blocks of Honey June corn. Outcrossed corn occurring in the Honey June variety was determined in the proximal middle and distal rows of each block. A tabulation giving the percentages of outcrossed grains in the rows at various distances from the borders of each of the Honey June blocks nearest the Yellow Surcropper variety in which these rows were at distances from the contaminant is given in table 152.

Table 152
Retardation of corn pollination by block
(data from Jones and Brooks, 1950)

Feet from block borders	Feet of the block borders from contaminant				
	0	82.5	412.5	990.0	1650.0
8.75	51.67	31.91	7.32	1.78	0.46
26.25	28.12	12.49	3.13	0.55	0.15
43.75	19.02	8.73	1.48	0.32	0.21
61.25	16.82	6.41	1.77	0.46	0.12
78.75	9.53	5.64	1.60	0.23	0.12

Blocks nearest the contaminant pollen source consistently had more out-crossed corn than those farther away. Rows of Honey June in blocks nearest the Yellow Surcropper also consistently had more outcrossed grains than those more distant. Distance effects were operative, there-fore, between and within the blocks of the Honey June variety. It is seen that percentages of outcrossed grains became lower at the longer distances within the Honey June blocks than between the blocks. Such reductions were believed by the authors attributable to the fact that the first few rows, those proximal to the contaminant, acted as hin-drances. Doubtless all rows acted as hindrances—more than the fields of stubble of small grain. Whether the corn plants functioned mechanic-ally, serving to sieve the pollen, or whether they possessed an attractant that terminated the dispersal journey is unknown.

Hindrances to passive disperser organisms must restrain or prohibit the agent that carries them. For example, on entering a forest, wind is decelerated. Particles or organisms borne by the wind must also become slower and eventually come to rest. A forest was found to be a hin-drance to dust particles, as shown by Geiger (1950) in table 153.

Table 153
Retardation of dust particles inside forests
(data from Geiger, 1950)

Distance from forest margin (meters)	25	50	100
Particles inside forest	14,000	11,800	1,500
Particles outside forest	10,300	10,200	10,100

Boll weevils, *Anthonomous grandis* Boh., entering uncultivated wood-land areas hibernate nearer the borders of cotton fields where trees, bushes and other objects impede their movements (Fye *et al.*, 1959).

Effects of corn, *Zea mays* L., rows as hindrances to crossing in cotton. *Gossypium hirsutum* L., were measured by Pope *et al.* (1944). Different numbers of rows of corn were employed to separate red-plant and green-plant colors, with the results given in Table 154.

Table 154
Retardation of cotton pollination by rows of corn
(data from Pope *et al.*, 1944)

Corn rows from red-plants (number)	0	3	6	9
Red-plants (percentages)	27.2	17.8	14.6	12.9

Different amounts of restriction to the crawling of a mealybug, *Pseudococcus njalensis* Laing, upon different surfaces were found by Cornwell (1956). Movements of the insects were measured over a 48-hour period as they crawled over surfaces of smooth paper, sand, and cacao (*Theobroma cacao* L.)—littered soil. Cacao-littered soil, the coarsest of the surfaces tested, retarded mealybugs more than the smoother surfaces, and retardation was greater with nymphs than with adults. These restrictions are indicated in table 155.

Table 155
Retardation of dispersion of a mealybug
(data from Cornwell, 1956)

Surfaces over which movements occurred

	Paper	Sand	Cacao litter
Av. no. mealybugs per plot	37	21	7
Av. adult/nymph ratio	3.51	0.87	0.45

Roughness of the course retarded both nymphal and adult stages. That nymphs were restricted more than adults is not an unexpected occurrence, since the nymphs are smaller in size. Plywood surfaces were less restrictive of movement than cacao-littered soil. Definitive data, given as mean number of mealybugs per plot over a 14-day period, are summarized in table 156.

Table 156
Retardation of cacao litter to mealybug dispersal by ants
(data from Cornwell, 1956)

Surface	Ants absent	Ants present	Average
Plywood	20.3	13.4	16.9
Cacao litter	11.0	7.5	9.3
Average	13.7	10.5	—

The searching capacity of a ladybeetle, *Stethorus picipes* Casey for mites was restricted by dust particles, according to Fleschner (1958). In 15 minutes, larvae of the ladybeetle travelled 259 inches on undusted paper, 53 inches on paper with 320 kapok dust particles per square inch, and 8 inches on paper with 1,600 dust particles per square inch. In the absence of dust particles the larvae are effective in controlling plant-feeding mites. In the presence of much dust, as on trees beside a much-travelled road, for example, mite control cannot be effective.

It was suggested by Cragg and Hobart (1955) that a sudden release of marked blowflies, *Lucilia sericata* (Mg.) and *L. caesar* (L.), for experimental purposes in a wild native population may seriously affect the distance of dispersal. It is not known whether the effect of a great and sudden increase in the population would be one of increase or decrease.

More green peach aphids, *Myzus persicae* (Sulz.), are frequently seen on plants enclosed under cover, such as shade tents of the type used for growing tobacco. Tent-covered plants, by virtue of nutritional factors or favorable temperature, moisture, or other conditions, may be productive of aphids more than open-grown or unshaded plants. Coverings may act as hindrances to predators and parasites of the aphids. Open-grown plants were found to be effective hindrances to the incidence of virus Y disease in potatoes, the causative factor of which is spread by aphids. This relationship is shown by data from Waggoner and Kring (1956) in table 157.

Table 157

Incidence of virus Y disease in open-grown and tent-grown plants
(data from Waggoner and Kring, 1956)
Percentage of plants infected at feet from the source

	Percentage of plants infected at feet from the source											
	0	*1*	*2*	*3*	*4*	*5*	*6*	*7*	*8*	*9*	*10*	*11*
Open-grown plants	100	25	8	0*								
Tent-grown plants	100	93	80	66	50	36	25	20	14	8	3	0

*Data taken from Waggoner and Kring (1956), fig. 1 after "smoothing" the data for the tent-grown tobacco plants.

Tsetse flies, *Glossina* sp., live in more abundance along streams and in shaded locations but fly into clearings to attack their host, Moggridge (1949). Open treeless areas were shown as hindrances to the flies, especially during the dry season. Clearing of undergrowth and low branching trees permit the passage of dry and hot winds to dissipate the essential riverine ecoclimate that affords favorable conditions for the flies (Nash and Steiner, 1957).

The "Itigi-type" thicket was a barrier to *Glossina morsitans* Wst., as shown by Swynnerton (1936). The number of flies penetrating the thicket following man is given in table 158.

Table 158

Tsetse flies follow man target into thicket
(data from Swynnerton, 1936)

Distance flies followed man (yds.)	50	500	1,000	2,000	4,000	6,000
Flies following man (no.)	27	19	18	10	2	1

A clearing was found by Jackowski (1954) to act as a hindrance to the mosquito, *Aedes polynesiensis* Marks. Clearings thus serve to reduce filariasis, a disease whose causal organism is carried to healthy individuals by the mosquito.

One or more barriers permit reinvasions of the Mediterranean fruit fly, *Ceratitis capitata* Wied., into areas along the Seine and Rhine rivers on occasions subsequent to their disappearance in winter. Favorable weather and other conditions, according to Baas (1959), allow continuous or spotty infestations to extend from reservoir populations of the fly in southern France.

Reinfestation of previously infested areas must be a common occurrence. Many factors are doubtless responsible for the destruction of a species in an area. Reinfestation by the Mediterranean fruit fly subsequent to severe winter or meteorological conditions was discussed above. It is an inference from the very comprehension of the term *reinfestation* that dispersion occurs. Although such movement may often be an unusual or restricted one, it is essentially a dispersal activity. In instances of an eradicatory effect, as one achieved by a pesticide treatment, discussed by Keitt and Palmiter (1937), Joyce (1956), and Edmunds and Blakeslee (1959) organisms disperse into the previously uninfested area. Volcanic action, described by Eggler (1959, 1963) may destroy a species of organism in an area where repopulation will be made from outside the area of a previous distribution.

Frequency of dispersal. Repeated and consistent efforts are made by individuals of species to move into adjoining areas, although such areas may not permit development and reproduction. Recurring efforts are doubtless made to breach barriers of unfavorable conditions and to enter areas where conditions are favorable. Quarantine regulations are based on the theory of such efforts. Occasionally barriers are breached by individuals through their own energy and "long-distance" dispersal, as it is termed by some writers, occurs. History is replete with instances of such frequencies of dispersal. Two instances of recently reported infrequent dispersal and successful introduction are discussed below.

Intercontinental spread of a banana leaf spot disease caused by the fungus *Mycosphaerella musicola* Leach was discussed by Stover (1962) as dispersed by wind or possibly by infested material that was moved into previously uninfested areas. It was suggested that the ascigerous stage of the organism provided "sexual recombination of genetic factors, *transportation over long distance by wind-blown spores* and surviving through periods unfavorable to conidial activity." The development and release of enormous numbers of spores in Australia followed by dispersal by means of fortuitous air movements and specific climatic conditions were theorized to have carried spores to Central America. Widespread and almost simultaneous epidemics of the new disease recognized in the Caribbean area were accepted as significant evidence for the distant dispersal of 15,000 miles, which may have required 37 days. The impingement of spores on banana foliage under suitable climatic conditions was necessary to provide successful introduction in the newly infested areas.

Intercontinental movement of the spotted alfalfa aphid, *Therioaphis maculata* (Buckton), probably occurred in the summer or fall of 1953, according to Dickson *et al.* (1955). It was considered to have originated in the Old World and to have terminated in New Mexico. Such dispersal may have occurred by successive steps, through cycles of reproduction as in the distribution or invasion of different areas. It seems more likely, however, that favorable conditions for dispersal were followed by equally favorable conditions at the termination.

Time, in greater or lesser amounts, is required for the dispersion of all organisms. Time is therefore a hindrance. This concept is accepted for all species, although some organisms disperse far more rapidly than others. Some phases, stages, or activities of a species may be expected to differ from others in dispersal time. Various organic and inorganic factors doubtless effect variations in the time spent in dispersal, as they do in distance.

More data are given by authors on the dispersal of organisms in relation to distance than to time. This may be indicative that dispersal distance is more important than dispersal time. It may be that more efforts in time, equipment, and material, however, are needed to determine dispersal time than to determine dispersal distance. The aging of organisms must also be a factor in dispersal and may be confused with the time involved; indeed, it may be impossible to determine between the two. (Aging is discussed on page 193.)

Cucurbit mosaic disease symptoms were found earlier by Doolittle and Walker (1925) on plants nearer to overwintering harborer plants, *Microamphelis lobata,* than on plants at greater distances. Time was reckoned following the first appearance of vector beetles, *Diabrotica vittata* (F.), on plants until mosaic symptoms appeared. Data from each of two years are given in table 159.

Table 159
**Incidence of cucurbit mosaic disease
based on days of exposure to inoculum
(data from Doolittle and Walker, 1925)**

Yards from Microamphelis lobata	*1*	*140*	*175*	*225*	*350*	*500*
Days to symptom appearance						
1920	17	36	61	53	43	48
1921	13	—	51	39	69	20
Average	15	36	56	46	56	34
Regression curve (Wolfenbarger, 1946) gave the following:	17	42	43	45	47	49

These data show that the proximity of healthy to diseased plants affected the time for the first appearance of symptoms. The computed regression curve suggests that at distances in excess of 140 yards there was little time difference of symptoms.

Movement of the crown-gall bacterium, *Agrobacterium tumefaciens* (C. F. Smith and Towsend) Conn., through segments of sunflower stems was studied by de Ropp (1948). Bacteria were placed on the upper cut surface of stem segments and cultured *in vitro.* Results were based on the frequency with which bacteria passed through the segments. A regression curve was drawn from the data by Wolfenbarger (1959) to show the percentage of tubes with bacterial colonies in relation to the interval of time between application to the stem segments and removal from agar. The regression curve shows a rapid rate of increase in the number of contaminated tubes to four hours, then a less rapid increase to twenty-four hours in a graph where there is poor agreement of observed and calculated values.

A fundamental concept of dispersal time appears in a study of *Paramoecium* sp. by Andrewartha (1961), who stated, "We know from the table of *t* (Fisher and Yates, 1948, bottom row) that half the population is likely to be no farther from the center than a distance equal to 0.6745 times the standard deviation." Calculations of time-distance movements, therefore, permit the data concerning dispersal in terms of time and of distance of *Paramoecium* given in table 160.

Table 160
**Time and distance movements of one-half
the population of *Paramoecium* sp.**
(data from Andrewartha, 1961)

Movement for one-half the population, hours	*1*	*2*	*3*	*4*	*5*	*6*	*24*
Movement for one-half the population to move, cm.	7.2	8.6	9.9	11.7	12.5	12.6	26.6

It was further calculated from the table *t* that only 1% of the population might be expected to exceed 2.5758 times the standard deviation. This is true when large numbers of cases are available to estimate the standard deviation, according to Wadley (personal correspondence). A computation, therefore, gave 101.5 cm. as the distance within which virtually all *Paramoecium* would be located after 24 hours of dispersal time.

Crayfish, *Cambarus alleni* Faxton, an active disperser species, moved more rapidly in larger than in smaller populations. Individuals in 5, 10, and 20 animal groups reached 400 cm. in 20, 15, and 10 minutes, respectively, according to Bovbjerg (1959). More coactions, or social stimuli (termed *kinesis*), was in a crowded area than in a sparsely populated area. Cognizance should be given to density in dispersal studies of disseminules, especially to those known as active dispersers.

It was reported by Cammack (1958) that "References to disease gradients in the literature are numerous but studies on the quantitative aspects of these gradients are rare." His further report, however, contained quantitative incidence data with reference to distance and time for infection and pustule manifestation of *Puccinia polysora* Underw. on corn, *Zea mays* L. Two 4½ acre plantings of corn were made, one in April and one in September, 1956. Potted infected corn plants were provided as the inoculum source at a point near the fields. A tabulation gave the data on percentages of infected plants and also in terms of *P. polysora* pustules per infected plant (table 161).

Table 161
**Pustules of *Puccinia polysora* on corn on days after introduction
of spores and at distances from point of introduction
(data from Cammack, 1958)**

First Planting

Days after spore producing plant introduction	Known meters from spore source			
	5	10	20	40
10	127	21	6	6
20	187	84	55	49
30	235	201	165	166

Second Planting

10	191	66	27	21
20	235	191	171	167
30	251	241	229	236

Average, Both Plantings

10	159	44	17	14
20	211	138	113	108
30	243	221	197	201

Two gradients are seen in the above tabulation, in which the number of pustules decreased with distance from spore source and increased with time after introduction of the spore-producing plants. Greater extremes were found at the longer distances than at the nearer distances, in both percentages of infected plants and pustules per infected plant. The author feared there was a secondary spread that flattened the curve because of outside sources of inoculum.

Estimated daily rates of dispersal of a grasshopper, *Mecostethus magister* by a mark-and-recapture procedure were reported by Nakamura *et al.* (1964). A range of 0.73 and 0.77 was found for movement from the 12 x 12 m.^{2} and of 0.21 and 0.23 m. for movement from the 84 x 69 m.^{2} quadrate. Dispersal was believed independent of population density. Rate of movement remained constant throughout the season, according to the authors.

Dispersal distance-time date of the six-spotted leafhopper, *Macrosteles divisus* (Uhl.), were given by Linn (1940). These data are shown in table 162.

Table 162
**Dispersal of the six-spotted leafhopper
(data from Linn, 1940)**

Data from text

Feet from source	30	150-300	400-500
Insects collected, days after release, number	4	12	17

Data from table 2

Yards from source	50	100	150	200
Days to ½ of insects collected Insects collected, total number	36	9	8	2

Somewhat similar rates of curvilinearity were shown in regression curves by Wolfenbarger (1946) for the above data.

Adults of the New Guinea sugarcane weevil, *Rhabdocnemis obscura* Bvd., moved downwind farther in less time than they moved upwind in more time, according to van Zwaluwenburg and Rosa (1940). This was found through field trapping, painting, releasing, then recovering as many beetles as possible. A low rate of recovery, 1.29%, of the released beetles was obtained, a lower percentage than is usual of released specimens. Differences between upwind and downwind dispersal were significant, but differences between sexes were considered of no significance. The data on sexes were combined, therefore, to show average distances and time engaged in movement upwind and downwind and are given in table 163.

Table 163
Dispersal of the sugarcane beetle borer upwind and downwind
(data from van Zwaluwenburg and Rosa, 1940)

Wind direction	*Feet moved*	*Days spent in movement*
Upwind	368	48
Downwind	456	34

Computations give 7.7 feet as the average movement distance upwind and 13.3 feet as the average movement distance downwind per day.

Dispersion of the New Guinea sugarcane weevil, *Rhabdocnemis obscura* Bvd., was studied as to distance and time by van Zwaluwenburg and Rosa (1940). Data are given in table 164.

Table 164
Upwind and downwind dispersal of the New Guinea sugarcane weevil
(data from van Zwaluwenburg and Rosa, 1940)

SSW direction (downwind)

Ft. from release site	305	365	620	695	740	805	825	985	1,050	1,210	1,240
No. beetles rocovered	72	50	11	11	1	16	4	0	3	2	1
No. days to recovery	24	46	41	45	41	53	24	—	38	55	82

NNE direction (upwind)

Ft. from release site	265	300	405	600	740	825
No. beetles recovered	30	10	30	3	4	1
No. days to recovery	52	50	47	97	69	'

Significant differences were found by the authors between "downwind," SSW, and "upwind," NNE, as to distance and days of dispersal. No significant difference, however, was found attributable to directional responses by sexes. Regression curves, drawn by Wolfenbarger (1946), showed recoveries of more beetles were higher initially downwind than upwind and more beetles dispersed more distantly downwind than upwind. Time dispersal was shorter downwind than upwind. A trend is seen in these data for an inverse relationship between distance and time of dispersal, as is shown in figure 39.

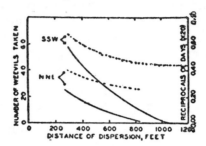

Fig. 39. Dispersion distance and dispersion time data for the New Guinea cane weevil from release to recovery points are shown for opposite directions from the release point. Dotted curves are reciprocals of dispersion-time; solid curves are for distance (data from van Zwaluwenburg and Rosa).

Data from four release-recovery type testing experiments by Dobzhansky and Wright (1943) were reduced to the average number of meters dispersed per *Drosophila pseudoobscura* fly per day. Although data were not given for any recovery in excess of five days for part of the experiments, the data from all were pooled to give the average accumulated distances dispersed through 8.5 days. Studies suggested that a semilogarithmic or a logarithmic-logarithmic function depicted the regression form of relationship of distance and days given for the data. Observed and calculated values are given in table 165.

Table 165
Days for dispersal of *D. pseudoobscura*
(data from Dobzhansky and Wright, 1943)

Days after release		1	2	3	4	5	6	7	8.5
Distance m./fly/day	Observed	53	62	74	81	84	104	75	89
	Log.-log.	55	65	72	77	82	86	89	93
	Semi-log.	53	66	73	79	83	86	89	92

Dispersal was most rapid immediately after release. Deceleration is evident thereafter, owing perhaps to aging (p. 193). Regression studies of data from individual experiments I, II, III, and IV showed that log.-log. or semi-log. transformations were about equally satisfactory.

Marked plum curculios released by Steiner and Worthley (1941) were later recaptured by jarring trees at distances from the release point; time (from release to recapture at distances) was also recorded. These data are given in table 166.

Table 166
**Numbers of and days to recapture plum curculios
at distances from the release site**
(data from Steiner and Worthley, 1941)

Feet from release to recapture	50	94	138	235	338	478	671
Number curculios recaptured	47	26	9	5	1	1	1
Number days to recapture	8	13	19	23	50	10	50

Most beetles were recaptured within 100 feet of the release site. Rather close agreement of the regression curve with the observed values was found by Wolfenbarger (1946) for these data. A curve drawn to show the days that elapsed between release and recapture is fairly close to the 50, 94, 136, and 235 feet observed distances, but the observations at 335, 478, and 671 foot distances depart rather distantly from the curve. These curves are in inverse relationships, as is shown in figure 40.

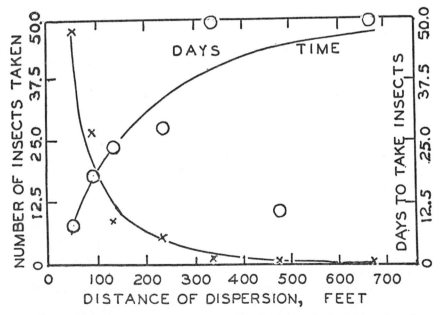

Fig. 40. Dispersion time and dispersion distance of plum curculios from release to recovery points. Bottom curve gives number of insects regressing on distance. Upper curve gives number of days spent in arriving at distances (from data given by Steiner and Worthley).

Data from experiment 2, given by Dobzhansky and Wright (1943), were used for drawing regression curves for daily recoveries of released *Drosophila pseudoobscura* Frowola. This is one of four similar experiments given and it appears to resemble the other three, as is shown by Wolfenbarger (1959). The top curve in figure 41 is the recoveries made the first day, and each lower curve represents observation on the second to sixth days.

All curves tend to converge at low levels of recoveries near 160 m. from the release point. Recoveries extend to distances in excess of the maximum observed and reach zero at some unknown distance. Equalization of the population over the area was a trend at the latest times (days) of recovery efforts.

DYNAMICS OF DISPERSING POPULATIONS

Many data were given on the time-distance dispersion of *Drosophila pseudoobscura* Frowola by Dobzhansky and Wright (1947). A tabulation was made to show the relationship of average number of flies re-

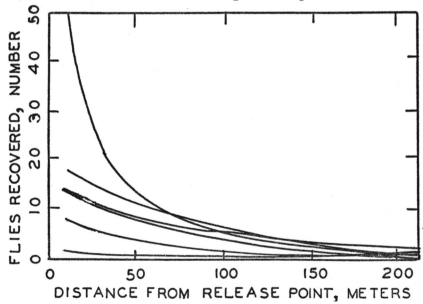

Fig. 41.   Dispersion of fruit flies with curves showing time recoveries. Upper and succeeding lower curves are first and successively later day recoveries, respectively. (Data from Dobzhansky and Wright, curves from Wolfenbarger, 1959).

captured in days after release at different distances. This is given in table 167.

Table 167

Recapture of *D. pseudoobscura* at distances
and on days after release
(data from Dobzhansky and Wright, 1947)

Meters from release point	Days after release							
	1	*2*	*3*	*4*	*5*	*6*	*7*	*8 & 9*
10	64.8	34.8	22.3	12.6	8.4	11.1	3.2	1.9
50	13.1	15.2	9.5	6.3	3.9	6.0	3.3	1.6
90	6.7	9.0	6.4	4.5	2.7	4.6	2.2	1.2
130	4.0	5.9	4.6	3.3	2.1	3.7	1.4	0.9
170	2.8	3.3	3.3	2.5	1.7	3.1	0.7	0.6
210	1.4	1.5	2.3	1.8	1.4	2.6	0.2	0.4

Two trends are seen in the above tabulation. First, the number of flies decreased with distance from the release site. Second, with increases in time there are tendencies for the number of recaptured flies to decrease at each distance from the release site. Recaptured fly numbers tended to reach low levels at 130, 170, and 210 m. on all days after release, however, so that numbers taken on the later days are more similar to those taken on the earlier days after release.

A very significant principle is recognized in the above tabulation. Although most flies were taken nearest to the release site the first day after release and (with the exception of the fifth day) continued to show such relationships, the numbers declined rapidly with time and distance. The numbers of flies recaptured tended to become and to remain low, as based on distance from and on time after release. Such decreases were explained by Wadley (1957) as attributable to (1) "spread to wider periphery" and (2) "drop out" of organisms. Many or all factors affect the decreases. Other terms, such as "spatial distribution" and "response to crowding," may be taken for (1). Changes to sedentary forms or to death may explain "drop out" in the terminal dispersion of previously active numbers of a dispersing population. This illustrates the dynamics of dispersing populations. It shows also the tendency to an equalization of populations over the distance range, which is also discussed later.

Definitive data on the distance and time of dispersal of the *Aedes albopictus* (Skuse) mosquito were given by Bonnet and Worcester (1946). They found through the release of laboratory-reared insects that the dispersal rate averaged 15.2 yards per day over a 21-day period and that the rate of movement decelerated. The rate of deceleration is illustrated in table 168.

Table 168
Yards dispersed per mosquito per day
(data from Bonnet and Worcester, 1946)

Days after release	2	7	12	17	20-21
Yards dispersed per mosquito per day	20	14	8	12	7

Although a regular regression curve formula was computed by Wolfenbarger (1959), yards dispersed $= 19.96 - [0.62\ (\times)]$, there is lack of close agreement of observed and calculated values. The curve suggests that most distances are covered immediately after the release and that shorter distances are traversed each day.

Release and recapture data on the tsetse fly, *Glossina morsitans* Wst., were reported by Jackson (1940). These data are given in table 169.

Table 169
Miles dispersed by tsetse flies
(data from Jackson, 1940)

Weeks after release	1	2	3	4	5	6	6.5
Miles dispersed, avg. accumulated	0.39	0.46	0.63	0.51	0.67	0.57	1.09

A regression curve computed from the above data by Wolfenbarger (1959) gave a "smooth" curve that may illustrate a principle. Data used for drawing the curve were used for calculating distances traversed per week, converted and expressed in feet in table 170.

Table 170
Dispersal of tsetse flies computed
from data in table 169

Days after release	1	2	3	4	5	6	6.5
Feet dispersed/week	1,742	845	528	370	264	211	106

Most dispersion occurred during the first week and (increasingly) less each succeeding week.

Male tsetse, *Glossina* sp., flies dispersed an average of 365 yards in one week, 242 in the second, 257 in the third, and 237 yards in the fourth week, according to Jackson (1946). Initial and terminal rates of dispersal of the rice weevil, *Calandra oryzae* L., as measured in time, were reported by Kono (1952) also to differ. Decelerated rates of dispersal may be attributed to space satisfaction of individuals, or to aging, or both.

All organisms need optimal or require minimum limits of space, both inter- and intra-specifically, for reproduction and development, according to Dice (1952). Part of organismal dispersion may be attributed to the seeking of optimum spacing between individuals. Dynamics of animal interactions or population actions, discussed by Thompson (1939) and Nicholson (1955), seem likely to influence or even govern dispersal. The successful presence of an organism in a location may be attributed in part to optimum, or at least to acceptable spacing in relation to other organisms. Timing or speed of movement would be expected to assist in optimal spacing. Some individuals would move or be moved more rapidly in a population than others.

Departure time from release site. Times for departure of dyed houseflies from release sites were given by Morris (1964) in data that are apparently unique for this activity. Data taken at several barns on zero and up to several days after release, averaged from pooled data, are given in table 171.

Table 171
Recovery data, days after release
(data from Morris, 1964)

Days after release	0	1	2	3	4	5	6	8	9	10	12	16	17	22*
Recovered flies (%)	41.8	32.9	19.1	11.0	9.0	20.4	7.4	10.0	1.9	4.9	2.6	2.0	1.0	0.7

*So few recoveries, less that 0.5%, were made on certain days, that they were omitted.

Steady decreases in the percentages of flies recovered following release are seen in the data obtained on days following releases, except for the fifth, eighth, and ninth days. Such exceptions are often found in extensive series of field data and, although they bear consideration, are lacking in significance. Decreases are most rapid within about the first six days and are less rapid from 6 to 22 days. (Transformation of days to logarithms tended to rectify the data for graphic comprehension, although the modified semilogarithmic method of regression curve computations described by Wadley and Wolfenbarger (1944) would likely provide a more functional relationship of the data.)

A consideration was given by Barber and Starnes (1949) to houseflies and to the time they spent in flight. Percentages of 4.3% and 16.2% were given for female and male, respectively. More time was spent with the increasing age of houseflies to about 7-9 days, after which the percentage of time spent in flight decreased.

Any introduction of a parasite into a new area presumes dispersal from the introduction site. It is expected and hoped from an economic viewpoint that the species will disperse in time over the distances in order to attain an objective. An example was given of increased parasitation of the European corn borer, *Ostrinia nubilalis* (Hbn.), by *Lydella stabulans grisescens* R. D. at a nearer distance in a shorter time than at a greater distance in a longer time. Data on percentages of annual parasitization of the European corn borer to seven years after release at two radii from the release site were given by Baker *et al.* (1949), as seen in table 172.

Table 172

Parasitization at two distances and seven years
after parasite release
(data from Baker *et al.*, 1949)

Within radii of parasite release site, miles	*Percentage of larvae parasitized in*						
	1932	*1933*	*1934*	*1935*	*1936*	*1937*	*1938*
3.5	0.3	2.8	6.3	7.6	10.0	17.1	20.9
7.5	—	—	—	4.4	7.0	9.6	13.0

These data show continued increases in parasitization from 1932 through 1938 at the 3.5 mile distance following the release of *L. stabulans grisescens* in 1927. These increases must approach an equilibrium that must have occurred after 1938, according to the above data. Equilibrium of parasitization (approaching total or perfect parasitization but seldom or never attaining it) must eventually be attained at all distances.

The data tabulated above as time spent on dispersal were converted to reciprocals and were drawn as regression curves. These curves may be compared as to slope and show similar rates of curvature. Decreases in number of weevils were larger than is indicated by the reciprocals for the time spent in dispersal. This suggests that the numbers of insects decrease rapidly and those that continue to disperse do so at less rapid rate.

A generalization or comparison of dispersion as measured for organisms which terminate at distances or in time was given by Wolfenbarger (1959). These comparisons considered regression curve slopes. Dispersion distance curves were characterized by having steeper slopes of decrease than reciprocals of dispersion time cures. Regression rates of decreases in numbers of insects with distance increases from the origin were more rapid than were regression rates of increase in time required for dispersion. This suggests that the numbers of insects decrease rapidly and that those which continue to move must do so at a less rapid rate. Much of the significance of this dispersion behavior characteristic, however, remains to be determined.

The speed of migration of the migratory butterfly, *Ascia monuste* L., was studied by Nielsen (1961). The speed varied, depending on wind direction and butterfly direction movements. Data from the author's table 4 given in km./hr. and the data in table 5, converted to km./hr. for 50% to pass, are given in table 173.

Table 173
Speed of the migratory butterfly, after marking
(data from Nielsen, 1961)

Distance from butterfly marking point, km.	*1.7*	*6.0*	*15.0*
Table 4	12.4	11.2	11.3
Table 5	15.6	15.0	18.7
Mean (or average) speed of dispersion, km.	14.0	13.1	15.0

Speeds reported in table 4 were consistent, and show that butterflies flew as rapidly to 15.0 as at 1.7 km. from the release site.

Experimental studies with the rice weevil, *Calandra oryzae* L., in which time lapse was discussed, were made by Kono (1952). Changes in the cumulative number of individuals disappearing from a container conformed to a formula:

$E(_t) + = (1-q^t)$, where
$E(_t)$ is the number of individuals lost in time t,
N is the number of individuals introduced originally
p is the probability of dispersion in unit time and
q is $1-p$

Speed of movement of the oriental fruit fly, *Dacus dorsalis* Hendel, when attracted to methyl eugenol was reported by Steiner (1952). A summary of the data is given in table 174.

Table 174
Time for oriental fruit flies to reach methyl eugenol
(data from Steiner, 1952)

Yards from methyl eugenol	*100*	*415*	*830*
Minutes for flies to reach attractant	2	6	16
Mean yards speed	50	69	52

Nearer flies were attracted to the lure more quickly than those at longer distances—as might be expected. Speeds of movement suggest that flies dispersed almost as rapidly to 830 as to 100 yards.

Loss of self-control. Control of the dispersal journey of active disperser organisms may be essential with many species. Loss of control, involuntary movement, or passive dispersal, on the contrary, may be essential to other species. Such loss of control implies passive dispersal. Some species may have active and passive dispersion, perhaps in different stages. Aphids—small, apparently buoyant insects—are often considered weak fliers, but various records show they disperse widely. It is not unexpected that aphids may occasionally disperse among lands of the Caribbean area, as suggested by Wolcott (1955). "Aphid showers" have been observed in south Florida, in which the insects, numbering perhaps three or four per square foot of surface, were present on plants, stones, boards, pools of water, and other places where they had terminated the previous dispersal journey. These showers could be accounted for in many cases by plant sources—often potato plants—on which aphids were developing in the area. In other instances the dispersal might have originated in one of the Caribbean lands or in a more northern area.

Aphids initiate flights at low (3 mph.) wind speeds, according to Davies (1936) and Thomas and Vevai (1940). Once airborne, however, aphids lose control over their flight direction at wind speeds of 1.2 to 1.5 mph., according to Haine (1955). Since air movements (winds) must usually exceed 1.5 mph., it may be hypothesized that aphids usually lose control of their dispersal journey and are moved passively. This may be an important part of the biology of many aphids. Actually, global dispersion of the worldwide distributed *Myzus persicae* (Sulzer) may recur periodically.Global dispersal of other aphid species may also recur.

Leafhoppers, discussed in chapter 12 as having continent-wide dispersal, may be very passive.

Much loss of life of organisms may occur during dispersal, as discussed by Morris and Miller (1954) for the spruce budworm. Some individuals may, however, terminate dispersal alive and in viable condition to reproduce the race.

If aphids and leafhoppers disperse globally, certain other insects may do it also. Distances are such effective barriers, however, that most organisms initiate, disperse, and terminate the journey not far from the origin.

Frequency or number of organisms that disperse globally may be a significant consideration. Such dispersion may be more likely accomplished through island- or continent-hopping rather than by a nonstop journey from initiation to termination. This would be through successive cycles of generations.

II
ORGANIC
FACTORS

7

External Factors

Living forces behaving as an organic whole or organization in a coordinated program or pattern of activity may be said to function organically. Host or food, density of the same and other species, enemies and sex are organic factors, contrasted with inorganic factors such as area or locality, directions, or barriers such as mountains or bodies of water.

Activations for dispersal may originate from within (internal) or without (external)an individual organism. Inorganic factors discussed above are usually those that are external in origin. External factors include (1) grouping of individuals, (2) effects of hosts and their densities, (3) inter-and intraspecies densities, (4) opposite sex, (5) enemies and (6) other factors. Internal factors include aging, nutrition, periodicity and mating.

It is often difficult or impossible to determine whether the predisposing factor, motivating principle or "urge," according to Nielson (1961), affecting dispersal is external or internal. Sometimes one factor is dominant, sometimes it is another sometimes two or more may be operative.

GROUPING

Grouping is manifested by species representing many orders and families of motile organisms. Such grouping, although a form of dispersal, may occur before or during the principal disperse phase. Terms that are somewhat synonymous with grouping include: aggregation, assemblage, cluster, collection, congregation, consortism, convergence, flock, and swarm. The act of grouping doubtless has some advantages and also some disadvantages to the species. Grouping is a requirement with the social insects—the honeybee, for example. Grouping provides for sexual selectivity among individuals. Dense populations may alter the substrate on or in which organisms live. Such change is needful or desirable in many instances, harmful in others. No known biological purpose of grouping of *Aedes taeniorhynchus* Wied. mosquito larvae, however, was recognized by Provost (1960). It seems likely that grouping sometimes provides protection and sometimes assists in food gathering. Predation may be more extensive among groups of organisms than among solitary living members. There must be, however, some fundamental need and also some cooperation among organisms in order to achieve grouping.

Grouping is generally assumed to consist of the nearness of an organism to the nearest neighbor in discussions by Dice (1952), Clark and Evans (1955), and Cottam and Curtis (1956). *A group may be defined as a collection of organisms in which each individual is more closely related to other members of the collection than to any individual outside of it.*

Studies of epidemiology may also be considered studies of grouping. An example is the study by Ederer *et al.* (1964), who also used the term *clustering* in reference to leukemia disease incidence.

Aggregation, clustering, grouping, or other contagious distributions may be attributed to various factors—some external in origin and some internal. These are accepted at the outset as nonrandom distributions. Some causes of grouping occur because of rapid multiplication; hence, the time factor enters into consideration. Another cause of grouping is a stimulus, external or internal, of some particular kind. Aggregation is a fundamental characteristic or behavior with the more highly specialized species, according to Waters (1959). It has a direct relationship with reproduction and survival, with ranges from sexual pairing of solitary living species to massing of highly organized social species. Low populations are believed random in distribution owing to scarcity. Medium population densities tend to aggregate. Individuals in dense populations tend toward randomness and may indicate saturation levels. *Drosophila* sp. were found by Dobzhansky and Pavan (1950) to form nuclei of high and low population densities. Some species were reported as tending to form discrete nuclei and others to be distributed relatively more uniformly.

Grouping of organisms ranges from the isolation of a few individuals to swarm numbers of social insects. Such grouping may be either an instinctive behavioral response or a selective action. Grouping may occur among active and passive disperser organisms.

One or several attractants may function with active disperser organisms and result in the assemblage or convergence of individuals or groups in or to one spot or area. Such attractants may be initial stimuli resulting in nonrandom distribution. Such distributions are generally recognized biological occurrences that may be termed *contagious distributions;* these have been discussed by Neyman (1939), Ashby (1948), Dice (1952), Beall and Rescia (1953), Goodall (1952), Greig-Smith (1957), Henson (1959), and others.

A discussion of grouping in terms of the *stochastic* process of studying populations treated them "as aggregrates of groups or as aggregates as clusters," according to Neyman and Scott (1959). A hypothetical example may be given as follows: A newly discovered infestation of the dreaded Mediterranean fruit fly, *Ceratitis capitata* Wied., was found in a grove. Questions immediately arose as to other infestations. Although other infestations in the region were unknown, there was greater likelihood of the discovery of an infestation in adjoining groves than in those at distances (10 miles) from the known infestation. A second factor of the stochastic process, according to Neyman and Scott (1959), is that providing "a unified theory of spatial as well as numerical changes in the population."

Many references discuss grouping of insects as affected by distances. Some present quantitative data illustrative of the effects, although most are qualitative in character.

Some fungi (*Phytophthora cinnamomi* Rand of the Phycomycetes is an example) possess motility of zoospores with an active dispersal phase. It might be expected that such an organism would behave somewhat as other organisms that disperse actively. Selectivity was demonstrated for *P. cinnamomi* in its attraction to avocado, *Persea americana* Mill, by clustering on or near the roots, as reported by Zentmyer (1961). Quantitative evidence was obtained and is presented in table 175.

Table 175

Attraction of avocado roots for *P. cinnamomi* zoospores
(data from Zentmyer, 1961)

Distance from roots (mm.)	Average number of zoospores in areas 0.5 mm. square	
	Avocado	*Citrus*
0-0.5	34.0	0.6
0.5-1.0	14.7	1.4
1.0-1.5	11.1	0.9
1.5-2.0	8.7	1.3
2.0-2.5	5.0	0.9
2.5-3.0	4.2	1.3

Some attractant (root exudates were suggested) may have exerted a concentration gradient that stimulated encysting with greatest frequency nearest the roots. A lack of zoospores about citrus (Mandarin orange) roots indicates a lack of attraction.

In studies of principles, methods, and organization of blast disease of rice, *Piricularia oryzae* Cav., Ono (1965) gave some data that are pertinent to the distance of spore dispersal. Different wind velocities separated spores from suspension in water and flew some of them to a slide 20 cm. downwind (table 176).

Table 176
Separation of rice blast spores from a spore suspension
(data from Ono, 1965)

Wind velocity (m./sec.)	2	3	4
Spores caught on slide (no.)	1.3	8.3	49.3

These data are suggestive of the effects of wind on the separation of spores from a suspension. Greatly increased numbers of spores were separated by the increase of wind from 3 to 5 m./sec. A further test of spore separation was conducted in which dispersal was made by spores flying away from falling drops of spore suspension (table 177).

Table 177
Spore density rate after falling specified distances
(data from Ono, 1965)

Falling distance (cm.)	0	10	30	60	90
Spore density after fall	100.0	87.6	68.4	50.4	44.6

One-half of the spores dispersed by a fall of 60+ cm.

Dispersing winged aphids appear to move or be moved aimlessly and without having been attracted to any particular spot. There is evidently great wastage of individuals of this and other families as they disperse from the host plant of their origin to colonize other plants. Several species of this family are the important vectors of viruses and as such are of more importance than as direct pests. Host plants may be populated by chance rather than by attraction, since aphids are not attracted to hosts from a distance (Kennedy, 1950). Aphid-infested plants are usually first observed as spotted, unequally populated areas in fields. Certain spots will be free of aphids and others will possess more or fewer aphids than other areas. Eddy currents may be important, according to Johnson (1950), in causing grouped infestations.

A coccinellid, *Semiadalia ll-notata* Schneider, collects in groups and hibernates on hills, to which some individuals must move longer distances than others, according to Hodek (1960). Living of solitary lives and searching for food is abandoned with the shorter days in August and later, followed eventually by dispersal to hibernation sites. Hibernation occurs annually in the same localities on conspicuous objects that are isolated in the landscape. Groups of the beetles were believed to attract other beetles, resulting in the formation of large masses.

Roebuck *et al.* (1947) found in trapping studies on the behavior of the adult click beetles, *Agrotes obscurus* L., *A. lineatus* L., and *A. sputator* L., that hay heaps on clipped grass were the most attractive trap types. Relative attractiveness of the various types of traps is shown in table 178.

Table 178
Attractiveness of grass for click beetles
(data from Roebuck *et al.*, 1947)

Type of trap	Total no. beetles May 13-July 4 Avg. per trap
Hay on clipped grass	180.2
Hay on clipped grass surrounded by 6 in. of bare soil	113.2
Hay on bare soil	114.7
Grass on clipped grass	122.0
Grass on clipped grass surrounded by 6 in. of bare soil	124.0
Grass on bare soil	84.6
A pit 4 in. deep with clipped grass bottom and filled with hay	56.5
A pit 4 in. deep with bare soil bottom and filled with hay	59.5
A pit 4 in. deep with clipped grass bottom and filled with grass	85.5
A pit 4 in. deep with bare soil bottom and filled with grass	42.0

Consideration given to the area over which influence was effective suggests that the most attractive type trap extended to longer distances than did the trap with the least attractiveness. An alternative explanation to greatest attraction might be the repellent nature of the less effective traps. The more likely explanation is relativity or degrees of attractancy.

Low populations of larvae and pupae of the Nantucket pine shoot moth. *Ryacionia frustrana* (Comst.), were found randomized by Waters (1959). Grouping became apparent in moderate populations, then changed to randomness as populations increased. Low populations were randomized owing to scarcity, and high populations tended to become randomized owing to a saturation level.

Traps used for attracting and retaining mosquitoes frequently employ light emanating from incandescent lamps as the attractant. However, traps may be operated without lights and presumably without any attractant quality. Catches of mosquitoes without lights are therefore presumed to be specimens at the trap site that are taken largely by chance. Collections of male and female *Aedes taeniorhynchus* (Wiedemann) taken by traps operating with and without lights, as reported by Provost (1952), show more sex differential in those traps without than in those with lights. Data given of the average number of A. *taeniorhynchus* taken per trap light and female/male ratios are presented in table 179.

Table 179

Traps with and without light attracting male and female mosquitoes
(data from Provost, 1952)

Trap with light		Trap without light		Female/male	
Female	*Male*	*Female*	*Male*	*Light*	*No light*
6928	1354	324	26	5.00	12.46

These data indicate that at the trap site males were attracted to the lighted traps more than females were.

Catch of the European corn borer moths by light traps was found related to contours in a corn field by Ficht and Heinton (1941). The average number of moths per trap on different contours is shown as follows: high contour—1326, intermediate contour—1312, low contour—919. These data indicate the convergence of moths to areas of higher contours.

Any means by which pest insects might be concentrated could provide an effective means of control. Electric light traps are often used to lure insects to some particular spot, perhaps to the death of those attracted. Such traps were used by Ficht *et al.* (1940) in an effort to control the European corn borer. They reported that "the moths normally travel against the wind" and found that the traps on the northeast border of corn fields were the first to be encountered. Data on average number of moths per trap were given as follows: eastern row—401, second row—329, third row—253 and western row—148. More infested plants were found in rows nearest lights, according to Hervey and Palm (1935). Data showing the magnitude of infestation are given in table 180.

Table 180

Border row plants are more heavily infested with
European corn borer larvae than internal rows
(data from Hervey and Palm, 1935)

Row number	1	2	3	5	6
Plants infested (%)	84.6	62.7	50.0	39.3	37.8

It was suggested that the moths may prefer the "twilight" conditions rather than plants in the darker locations; hence, they infested most those plants nearest the lights.

The female sex lure of the gypsy moth, *Porthetria dispar* (L), has been used for more than 60 years to detect the presence of the species, according to Holbrook (1960). It was insufficiently attractive, however, to provide satisfactory control measures. Abdominal tips of the females were used for attracting males for the purpose of determining the presence or absence of the species. Although two tips attracted moths, successively more tips per location attracted successively more moths. This is shown by the data in table 181.

Table 181
**Multiplicity of tips of gypsy moth females attract more males
(data from Holbrook, 1960)**

Tips per trap (no.)	2	4	8	16	32
Moths caught (no.)	16.8	46.3	53.0	72.5	77.3

Although lure units were doubled successively, the numbers of moths caught were not doubled in like manner, except for the 2 to 4 tips per trap. It is possible that either the 2 or 4 tips per trap (or both) are out of line as to the number of moths caught. The above data show, however, that 8 or more tips trapped comparatively fewer than 4 tips. Comparatively less distance is traversed for 8 than for 4 tips by assuming that (1) equal moth populations were prevalent about all traps. (2) traps were placed to exclude influencing collections at other traps, and (3) conditions were equal over the trapping area.

In describing a technique for measuring the trapping efficiency of electric insect traps, Hartstack *et al.* (1968) showed more moths were taken nearest the light (table 182).

Table 182
**Adjusted number of moths in pans
at distances from the light
(data from Hartstack, *et al.*, 1968)**

Radius (feet)	*Moths (mean nightly catch)*
3	44.12
6	7.37
12	2.79
24	1.58
36	1.07
48	0.58

A very strong gradient is given, indicating an attraction for lights; as distances increased away from the lights fewer moths were caught. At 48 feet a much lower mean catch was obtained as compared with 3 feet.

Nocturnally active malarial mosquitoes, *Anopheles quadrimaculatus* Say, rest during daylight hours in darkened areas to which they are attracted with the coming of daylight. The use of such attractions was employed by Huffaker and Back (1945) in dispersion studies. Darkened enclosures, barns, and sheds, and nail kegs, were used in sampling mosquito abundances at distances from breeding sources. Although many more specimens were taken from barns and sheds than from nail kegs, the relative densities indicated that there was comparatively equal attraction to each convergence-type unit. This is shown in table 183.

Table 183

Relative density of malarial mosquitoes in barn or shed or nail keg
(data from Huffaker and Back, 1945)

Distance from source (mi.)	Relative density (number of 0.1 mi. taken as 100)	
	Barn or shed	Nail keg
0.1	100.0	100.0
0.5	50.0	31.6
1.0	5.0	4.2
1.5	0.2	2.1
2.0	Less than 0.1%	
3.0	Less than 0.1%	

Regression curves drawn by the authors also indicate relative attractancy of the two types of enclosures.

Influences of feeding and breeding areas are subject to question as affecting distances of dispersal of organisms. Two classical references on the dispersion of the housefly, *Musca domestica* L., are rather definite in answering this question. According to Parker (1916), favorable "breeding and feeding areas are not necessarily areas which attract flies and retain them, but that they may be considered substations, so to speak, which aid and abet distribution and further increase the final radius of dispersion." It was reported by Bishopp and Laake (1921) that "many apparently favorable feeding and breeding places were passed in the course of migration. . . ."

Selection of a release site dependent on the possible attractiveness or lack of attractiveness to houseflies, *Musca domestica* L., was made by Schoof and Siverly (1954), who provided data on dispersal distances at each site (see figs. 14 and 17). Tagged flies were taken at distances to nearly eight miles, the maximum distance from the release site at which traps were placed. The highest percentages of tagged flies were taken at the "hog farm" release site, and the lowest percentage was taken at the "rendering plant" release site. All regression curves have similar curvilinearity, indicating similar rates of housefly dispersion regardless of release site characteristics.

Adults of the narcissus bulb fly, *Merodon equestris* (F.), feed on the pollen and nectar of various flowers outside of bulb plantings; hence, they are attracted to weedy growth. Marginal influences were found in that more eggs were placed on plants nearer bulb field edges. This is shown from Doucette *et al.* (1942) in table 184.

Table 184
**Egg deposition of the narcissus bulb fly proximal to field margins
(data from Doucette *et al.,* 1942)**

Yards from woods	7	12	35	85	150	200	220	245	255	270	285	300
Infestation (% of plants)	37	73	33	15	25	19	18	4	14	11	3	4

Considerable reduction occurred within 85 yards of the field margins, although distances in excess of 285 yards appear necessary for low infestations.

Estivation assemblies of the bogong moth, *Agrotis infusa* (Boisd.), were found to occur in the same spots annually. Moths assemble at altitudes above about 4,000 feet in the Australian Alps, according to Common (1954), where concentrations of moths are largely associated with granite outcrops and isolated granite boulders on numerous mountain peaks. The moths migrate from the assembly sites in late summer and fall and return to the breeding grounds.

Host density and quality. Density of plant populations influences the incidences of plant diseases, according to van der Plank (1948). Such influence may be greater with virus diseases than with fungal or bacterial diseases. Hence, when systemic inocula enter the crop randomly and the proportion of infection is low, the number of infected plants per unit area is constant and tends to be independent of the density of stand. This is shown by data from Linford (1943), in which pineapple yellow spot was carried by onion thrips, *Thrips tabaci* L. (table 185).

Table 185
**Density of pineapples with incidence of pineapple yellow spot
disease, as percentage and number of diseased plants per acre
(data from Linford, 1943)**

Plants per acre	21,780	18,150	14,520
Infected plants (%)	3.3	4.6	5.2
Infected plants (no.)	718	829	755

Insect populations are often influenced by plant density. Although such populations are influenced by ecological situations, dispersion is also influential. Collard plants, *Brassica oleracea* L., were infested with more green peach aphids, *Myzus persicae* (Sulz.), per unit area of leaf surface in sparse than in dense plantings. This is shown in table 186.

Table 186
Green peach aphids per unit area of leaf surface with different numbers of plants
(data from Pimental, 1961)

Number of plants/7,500 sq. ft.	*30,600*	*1768*	*80*
Aphids per 20,000 sq. ft. in of leaf	20.2	21.4	57.3

Dispersal of mites to distances from carnation shoots and tomato plants differed but slightly, according to Hussey and Parr (1963). Data given by them are reproduced in table 187.

Table 187
Percentage of carnation shoots and tomato plants infested with mites at distances from the source
(data from Hussey and Parr, 1963)

Distance of dispersal (cm.)		*10*	*20*	*30*
Plant source of mites:	carnation (%)	63.8	22.9	13.3
	tomato (%)	64.6	16.1	19.3

These data suggest that the plant source of mites influenced distance of dispersal little or, probably, not at all.

Although insect density affects the distances to which species, stages, or phases of organisms disperse and the phenomenon is frequently observed, data are not often given. Qualitative results are given occasionally, however, such as those by Anonymous ("Our Military Expert") (1917) concerning mosquitoes and malaria. He reported that "the maximum flight from a profusely producing breeding place was one mile, while from places producing less abundantly the flight did not exceed half a mile."

Multiple-point releases of the yellow fever mosquito, *Aedes aegypti* (L.) were made by Morlan and Hayes (1958) to measure attack rates and determine households entered by mosquitoes. These data are summarized in table 188.

Table 188
Yellow mosquito attack rates and percentages of households entered from different numbers of mosquitoes released
(data from Morlan and Hayes, 1958)

Females released (no.)	*400*	*900*	*1,200*
Attack rates	57	90	187
Households entered (%)	62	58	56

These data were taken within a single block of release points. They show that attack rates increased as release numbers increased. Percentages of households entered showed slight decreases with release number increase. This may be significant in showing that no greater percentage of households could be entered by the dispersing mosquitoes and that they were forced to move farther.

Two behavioral features stand out from the studies by Bovbjerg (1960) of a Pacific shore crab, *Pachygrapsus crassipes* Randall, an active disperser organism. In these studies populations of different numbers were released in pools to react. Some individuals remained as residents at the release site, others dispersed. Data of these reactions are given in table 189.

Table 189

**Number of Pacific shore crabs released in different numbers
with portions dispersing and remaining**
(data from Bovbjerg, 1960)

No. at release site	5	10	15	20
No. remaining as residents (avg.)	2.5	3.0	3.5	4.0
No. of dispersants	2.5	7.0	11.5	16.0
Percentages that dispersed	50	70	77	80

Population increases at the release site gave rise to slightly increased numbers remaining as residents but to greatly increased numbers of dispersants. Percentage terms may be considered in which a gradient shows effects of population dynamics. A very rapid rate in the percentage increases was indicated from 5 to 20 at the release site. Such increases decelerated, however, with approach to 100 percent, and doubtless must be expected. Some individuals would be expected to remain at the release site. This is a very significant behavioral feature of populations.

Invasion of the cotton producing areas of the United States by the cotton boll weevil, *Anthonomous grandis* Boh., occurred by successive stages (Fig. 8). This may be attributed, in part, to the attractancy of cotton plants to the weevil. Part of the advance of the weevil into previously uninfested areas has been described by Hunter (1910):

"One of the primary reasons for the dispersion movement of the weevil is its inclination to obtain fresh food and cotton squares in which to breed. Where cotton fields are small and separated by considerable distance this instinct causes weevils to fly over a large extent of territory. Where fields are numerous it is unnecessary for much advance."

Kligler (1924) believed that human habitations restricted movements of *Anopheles* mosquitoes. Kligler wrote, "under constant wind conditions the distance of spread of *Anopheles* from their breed-place varies directly with the intensity of breeding and inversely with the density of settlements in the vicinity of the breeding-place." An observation by de Zulueta (1950) indicated that the number of adult mosquitoes captured is related to the size (quantity) of bait used as the attractant. Recaptures of mosquitoes, *Anopheles gambiae* Giles, showed grouping effects (nonrandom dispersion) in which the distances were related primarily to the distribution of human settlements (Gillies, 1961).

Varietal differences of receptors. Incidences of virus and other diseases, dispersion of insects, and incidences of plant pest injuries depend on host plants involved. Such host plants include species and varieties within species. Incidences of sugar beet savoy disease vectored by a lacebug, *Piesma cinerea* Say., were so related. Certain varieties were found by Coons *et al.* (1958) to show consistently less savoy than other varieties. Varietal responses are indicated by regression curves in figure 42. Rather wide divergences are seen in variety 215x216, which possessed the flattest and the lowest curve of disease incidence. Variety 555701-00 possessed the highest incidental curve, which also was flat. Variety 501712-00 possessed the most disease initially and the lowest incidence terminally of all varieties. Although preference by the vector insects doubtless accounts for most of the differences, other and perhaps unknown factors may also affect varietal incidences.

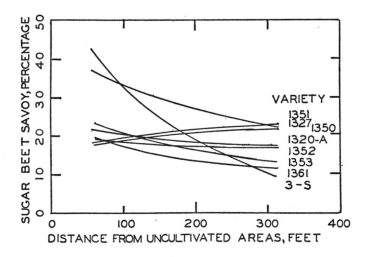

Fig. 42.   Regression curves of sugar beet savoy infections from different varieties of sugar beets (data from Coons *et al.*).

DISPERSAL DISTANCES EFFECTS OF MANIFOLD NUMBERS

More individuals would be expected to disperse or be dispersed far-ther from the source of a very large population, 100,000 for example, than from a very small population, such as 100. Some discussion is given of the comparison of such manifold numbers of small organisms at an origin and how this factor might affect the distances to which members of the population disperse. Several factors, such as chance occurrence, crowding, nutritional needs, and protection would be expected to affect dispersal distances. This section is devoted to the amounts or degrees to which the numbers of organisms are affected by distances. Although more research on the subject is needed, some definitive data and per-formance records are published and are here considered.

Manifold numbers (defined as many of one kind combined in the act of dispersing or receiving) may be organisms of the same species in the act of moving, or they may be receptive masses of individuals of the same or different species. Performance records are also essential to understand the behavior of manifold numbers and to indicate what may be expected. Some literature records are available, but some of these lack replication and uniformity of conditions that are conducive to much confidence.

Dispersing materials or organisms. Information on dispersal distances of different densities of organisms is desirable, since there are few such studies. A number of studies show, in general, that some organisms dis-perse farther from a large than from a small population. Doubling, trebling, or otherwise multiplying a population, however, gives less mul-tiplication of radial distances dispersed according to available data.

Water droplet dispersal studies reported by Faulwetter (1917b) in-dicated how drop sizes were related to the distance of spattered droplets (chapter 4 above). Consideration of the actions of different sizes of water drops may be somewhat analagous to densities of organisms. Spatter distance of droplets from a large drop of water would be ex-pected to exceed that from a small drop. Distance of droplet spatter from a large drop would be expected to be greater from a four- than from a one-foot fall. Data from Faulwetter (1917) from height-of-fall classes were pooled by drop sizes for presentation. Wolfenbarger (1946) took drop sizes of 0.04, 0.06, and 0.10 ml. as two-, three-, and five-fold larger drops than 0.02 ml. drops for drawing regression curves. Loga-rithmic conversions of drop size and distances of spatter were found to convert the data to straight line relationships. Data, as increased distance over the smallest drop size, for glass and blotter paper receptors are given in table 190

Table 190

**Dispersal distances of larger drop sizes on glass
or blotter paper impingement surfaces
(data from Faulwetter, 1917b)**

Impingement surface of drop	*Manifold greater drop size*		
	2	2½	5
Glass	22 (0.22) °	42 (0.42)	86 (0.86)
Blotter paper	20 (0.19)	42 (0.43)	72 (0.52)

°Fractional displacement increased distances over smallest size.

Increased distances of spatter were found for increased droplet sizes falling on glass or blotter paper receptors. Larger drop size at the origin increased the maximum distance of spattered droplets in each group and the increases were similar, except at the five times greater drop size on the blotter.

Experiments on splash dispersal of water droplet and of fungus spores dispersed by water were reported by Gregory *et al.* (1959). Data for distance classes to 65-75 cm. from target and drop diameters of 2, 3, 4, 5 mm. drops and for distances to 25-35 cm. from target and are given in table 191.

Table 191

**Splash distances of different diameters of drops
and heights of fall
(data from Gregory *et al.*, 1951)**

Drop diam. (mm.)	*Distances from drop target (cm.)*											
	2-3	3-4	4-5	5-6	6-7	7-8	8-9	9-10	10-17½	17½-25	25-35	*Total*
Height of fall 2.9 m.												
3	3.8	3.4	2.8	1.6	1.0	1.1	1.0	0.5	0.3	0.1	0.0	14.6
4	7.5	5.5	5.1	3.7	2.1	1.8	1.3	1.0	0.4	0.2	0.1	28.6
5	12.3	8.4	6.5	6.3	5.2	3.7	2.5	1.8	0.8	0.7	0.4	48.2
Total	23.6	17.3	14.4	11.6	8.3	6.6	4.8	3.3	1.5	1.0	0.5	
Height of fall 7.4 m.												
3	2.7	2.2	2.0	1.0	1.3	0.9	0.9	0.5	0.2	0.3	0.1	13.7
4	19.2	13.4	8.2	8.3	7.2	5.0	5.3	2.4	1.2	0.5	0.3	70.7
5	14.8	13.2	9.8	12.1	8.9	7.1	16.9	3.3	1.3	0.8	0.1	72.2
Total	36.7	28.8	20.0	21.4	17.4	13.0	13.1	6.2	2.7	1.6	0.5	

Gradients were found by the investigators in which (1) as distance from target increased the number of droplets decreased and (2) as drop diameter increased the number of droplets increased. More droplets were found from a 7.4 than from a 2.9 m. fall, except that the 3 mm. diameter drop from 7.4 m. was the least of any treatment.

In epidemiological investigations of maize rust, *Puccinia sorgi* Schw., Zogg (1949) observed that the distance of spread was associated with the density or abundance of spore reservoirs. Distance of dispersal was considered predictable for any size of spore reservoir.

DISPERSION: A FUNCTION OF DENSITY

Inoculum of wheat stem rust, *Puccinia graminis* var. *tritici* (Erikss. & E. Henn) Guyot, spore masses and its relation to infection of wheat was studied by Peterson (1959). Uredospores were dispersed over plants in a settling tower to fall on 10-day old wheat seedlings. Data were taken on "uredospores, germ tubes, appressoria, infection foci, chloronemic specks, and uredia per cm.^{2} of leaf surface." A regression curve was drawn by the author to show infection foci/cm.^{2}($x10^2$) and is reproduced in figure 43. It shows that as the number of viable uredospores increased, the number of infection foci also increased. Although a straight-line relationship curve was drawn and its regression coefficient was 0.331, an S-shaped curve appeared to give a more exact function of infection foci on uredospore mass. Means of three experiments of massed spores and the resultant cloronemic areas were calculated and give the results in table 192.

Table 192

Manifold wheat stem rust uredospores settling on wheat stem seedlings with chloronemic areas (data from Peterson, 1959)

Uredospores/cm.^{2}($x10^2$)	9.2	16.1	26.6	36.4	53.4	97.4	192.8
Chloronemic areas/cm.^{2}($x10^2$)	5.6	7.5	18.6	17.6	35.9	41.2	41.7

Disease incidence was closely related with inoculum density in figure 43 and in the above table. Although chloronemic areas increased as the inoculating masses increased, the increases diminished with mass increases, until at 97.4 uredospores/cm.^{2}($x10^2$) a doubling of the mass gave very little increase in resultant chloronemic areas. An illustration of the law of diminishing returns is thus presented.

A study was conducted by Bateman (1947a) in which 8, 4, 2, and 1 contaminating Icicle variety turnip plants were located in the central parts of plantings of Red Scarlet variety plants to determine contamination effects. Numbers of Red Scarlet variety seed were planted to provide proportional numbers to the central contaminating variety. Density in this experiment had no recognized relative or comparative effect

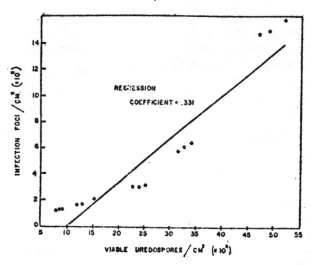

Fig. 43.   Regression of wheat stem rust infection foci on density of uredo-spores (fig. 7 reproduced from Peterson).

on the spread of contamination. Any effect of disproportionate numbers of plants surrounding the central contaminating variety is unknown.

European corn borer larvae, *Ostrinia nubilalis* (Hbn.), may disperse to some distance from the stalk on which the egg masses were placed. An excellent example of mass density is illustrated in data given by Neis-wander and Savage (1931), although the distance range was given in terms of "corn rows" (probably near three feet). A summary of these data as number of larvae is given in table 193.

Table 193
Larval dispersion from manifold egg concentrations
(data from Neiswander and Savage, 1931)

Eggs on central stalk	Distance to recover larvae (corn rows)			
	1	2	3	4
200	42	10	0	0
500	73	20	1	0
1000	115	57	7	1

According to the authors the "amount of dispersion increased as the population at the source increased, competition for food and space no doubt being the regulating factor. The distance to which larvae spread also increased with the population at the source. This was probably due to the increased chance of going long distance that comes with large numbers." Regression curves drawn for different larval densities by Wolfenbarger (1946) were found to reach zero at 3.1, 3.9, and 4.4 corn rows distances for 200, 500, and 1,000 eggs, respectively. These results show that doubling the population did not double the distance dispersed.

Studies of density and dispersion of caddisfly larvae of the genus *Cheumatopsyche* involved densities of 5, 10 and 20 larvae per bowl of 7 cm. diameter with rings of 3, 5 and 7 cm. Eight pebbles between 5.5 and 7.5 mm. in size were placed in some bowls as refuge for the larvae. Results, by Glass and Bovbjerg (1969), from the authors' Fig. 2 bar graphs are rather approximate and are given in table 194.

Table 194
Positioning of caddisfly larvae in rings in dishes,
with and without pebbles, in percentages
(from Glass and Bovbjerg, 1969)

Pebbles	Ring	Number of larvae per dish (%)		
		5	10	20
	Inner	88	70	49
Present	Intermediate	12	33	22
	Outer	3	18	24
	Inner	3	2	4
Absent	Intermediate	8	30	18
	Outer	85	83	75

Pebbles were retardants of movement of larvae to the outer ring, apparently affording protection for the larvae but functioning as barriers to dispersion. With pebbles absent most larvae moved distantly as far as possible, tending to desert the inner circle.

Dispersion is a function of density, according to Glass and Bovbjerg (1969). They supported the hypothesis that "animals demonstrating intra-specific agressive behavior, territoriality or spacing will disperse in a density-related fashion."

Comparison of larval dispersion from egg masses of 5,000 and 10,000 *Hippelates collusor* (Towns.) was made by Legner and Olton (1969), measured in terms of adult emergence. Rows of millet was the medium in which the larvae dispersed and developed, with results given in table 195.

Table 195
Emerging adults of *H. collusor* from
different egg densities at distances
(data from Legner and Olton, 1969)

Egg mass density	Row distance (cm.)		
	1 (30.5)	2 (42.2)	3 (53.9)
5,000	258.5	100.0	27.5
10,000	239.0	83.0	93.0

Rather similar numbers of emergent *H. collusor* emerged from each row (distance class), although actual numbers were greater (356:315) from 5,000 than from 10,000 eggs. The system was saturated by 5,000, according to the authors, to account for the numbers of emergent flies.

Abdomen tips of gypsy moth females were placed as lures for gypsy moths and results were measured by Holbrook (1960). Lure traps placed at 0.2, 6, and 12 feet caught averages of 27.0, 22.0, and 34.5 and 15.5 moths per trap, respectively. Different numbers of female moth abdomens, as bait strengths, per trap were compared with data in table 196.

Table 196

Male moths caught by manifold female moth abdomen tips
(data from Holbrook, 1960)

No. female moth abdomen tips	2	4	8	12	16	32
No. moths caught, 1954	16.8	46.3	53.0	–	72.5	77.3
No. moths caught, 1955	–	–	25.5	42.3	44.3	–

Increases in numbers of moths caught was found with increases in numbers of female moth tips used as lures. These increases were rectified fairly well through computing a regression curve whose independent variable, abdomen tips of female moths, had been transformed to reciprocals. The 1954 results showed greatly increased catches of 4 over 2 and but slight increases of 32 over 16 abdomen tips. A more rapid increase was found for both experiments between 8 and 16 abdomen tips. A consideration of attractants shows that two factors are involved. First, there is some distance from attractant to the organism; second, response by the organism must be made if a reaction is recognized.

Azuki bean weevil, *Callosobruchus chinensis* (L.), were released on a table by Watanabe *et al.* (1952), who showed that distance dispersed was affected by different numbers of individuals released at a center. Means computed from three individual lots of 50, two of 100 and one from each of 200- and 600-individual lots, over a rather limited time period of about 20-25 minutes, are summarized in table 197.

Table 197

Dispersal of Azuki bean weevils from manifold numbers released
(data from Watanabe *et al.*, 1952)

Weevils released (no.)	50	100	200	600
Dispersal distance (cm.)	8.6	8.9	9.5	10.1

Slightly longer distances were dispersed by manifold increases at the release site.

In a comparison devoted to different population densities of four species of insects, the green peach aphid, *Myzus persicae* (Sulzer); adult sweet potato weevil, *Cylas formicarius elegantulus* (Summers); fifth stage larvae of the Caribbean fruit fly, *Anastrepha suspensa* Loew; and three different sizes of the tobacco budworm *Heliothis virescens* (F.) were studied by Wolfenbarger *et al.* ('1974, in press) as to distances of movement. Significant differences were found for distances traveled between population levels for these insects through the use of simple mathematical regression computations. Regression of population size on distance was found to function logarithmically so that mean dispersal distances were further from the origin with population increases. The multiple regression equation for the four populations of green peach aphids, 15, 150, 1,500 and 15,000 (fig. 44), where Y is the expected number and X is distance =

$$\log_{10} Y = 5.7116 - 3.0113\ (\log_{10} X) + 0.6549\ (\log_{10} \text{population size}),$$

and it was significant at the 1% level of probability. The mean distances of dispersal were statistically significant, and the curve slopes appear characteristic for populations in that with greater populations zero density tends to be reached at greater distances than with lesser densities. These observations of population dynamics emphasize the importance of sparse and of dense populations. Pest populations reduced to sparse ones will be less important ones, injuriously. Beneficial populations—pollinators, parasites, and predators of pests—in dense populations will disperse more widely and be more effectively advantageous to man than sparse ones.

Multiplicity of honeybee colonies per acre was shown by Pankiw *et al.* (1956) to give more activities than low numbers of colonies per unit area. Although honeybees make no response to manmade plot borders, they may be placed by man so that distance barriers are effective to some degree as plot boundaries. Summarization of the data are given in table 198.

Table 198

Honeybee populations per square yard, number of tripped flowers, alfalfa seed per acre and honey per colony from numbers of colonies per acre
(data from Pankiw *et al.*, 1956)

Colonies per acre (no.)	0	1	3	5
Honeybees/sq. yd.	—	0.24	0.34	0.90
Tripped flowers (no.)	57	74	81	116
Alfalfa seed yield (lbs./A.)	35	45	72	117
Honey (lbs./colony)		113	92	73

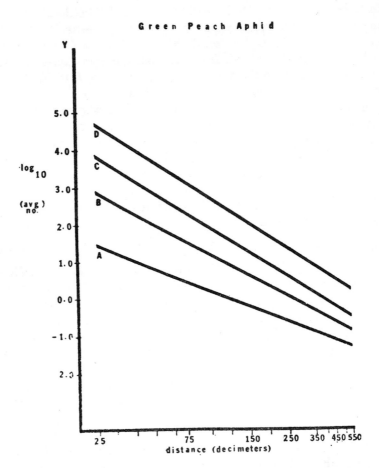

Green Peach Aphid

Fig. 44.   Regression of distance (decimeters) traveled by different populations of apterous green peach aphids released from a focal origin.

Dispersion of more bees from more colonies per acre gave larger
numbers of pollinators per unit area, more freshly tripped flowers, more
seed (yield) per acre, and less honey. To the experienced apiarist these
are but expected results. Although trebling or quintupling population
at an origin does not treble or quintuple the work accomplished, it gives
increase in work and decrease in seed. Beekeepers recognize that increasing pollination may decrease the honey produced per colony, owing
perhaps to competition for the available nectar by the multplicity of
gatherers.

Increasing the numbers of colonies of honeybees per unit of area has resulted in increased yields of seed and of squash, according to several workers. A summary of the results is given in table 199.

Table 199

Increased seed or squash yield from multiple numbers of colonies
(data from Pankiw *et al.*, 1956)

Reference	Crop plant	Colonies per acre (no.)						
		0	½	¾	1	2	3	5
Pankiw *et al.* (1956)	Alfalfa*	35	–	–	45	–	72	117
Blake (1958)	Crimson clover	–	–	384	–	509	–	535
Pankiw & Elliott 1959)	Alsike	127	–	–	218	–	220	–
Wolfenbarger (1961)	Squash	145	155	–	161	168	173	–

*Repeated from above tabulation.

Ascending orders of increase are noted in every instance. Increased amounts of bee colonies per acre were dynamics of population that induced pollination and increased yields of alfalfa, crimson and alsike clover, and squash fruit.

MANIFOLD RECEPTORS

Anything on which organisms impinge may be considered a receptor. Plants receive pollen, inocula of viruses, bacteria and fungi causing diseases, and infestations of insects or other pests. Insects may be receptors and may be nourished, attacked, parasitized, or mated with other insects. Some receptors consume the organisms, some add to the economy of the species, and others are neutral.

Receptor magnitude, target size or area, could be expected to affect the results of dispersion of organisms. In discussions on *crowd diseases* (defined as those caused by inocula that have dispersion of short distance ranges and have lives of short duration in the soil) van der Plank (1947a, 1948, 1949) emphasized the significance of size of fields. Larger and fewer fields suffered less loss from plant diseases than smaller and many fields. Specific diseases—groundnut rosette, and maize streak diseases and aphid-borne viruses of beet—were named. Inocula of crowd diseases, dispersing or being dispersed from outside a field, would be much less on a per-acre basis in large than in small fields.

Incidence of yellow virus in sugar beet crops was reduced by increasing the plant population, according to Blencowe and Tinsley (1951). Data are presented in table 200.

Table 200
Incidence of yellows virus disease (percentages) in sugar beets
by increased plant populations
(Blencowe and Tinsley, 1951)

| Year of observation | Plant row separation distance | |
	15 inches	30 inches
1946	12.1	20.0
1947	63.1	82.7

Although twice as many plants were present at the 15-inch spacing as
at 30 inches, there were 61 and 77 percent as many diseased plants in
these rows as with one-half the plants.

Influence of different plant populations per acre of pineapple plants,
receptors, were found related to the percentage of pineapple yellow
spot diseased plants by Lindroth (1953). Transmission of the virus inoculum was reportedly achieved by *Thrips tabaci* Lind. in random
manner. A tabulation of the data is given in table 201.

Table 201
Incidence of pineapple yellow spot disease in number
and percentage of infected plants
(data from Lindroth, 1953)

Plant spacing (inches)	12	15	18
Plants per acre (no.)	21,780	18,150	14,520
Infected plants (percentage)	3.3	4.6	5.2
Infected plants (no./acre)	718	829	755

Percentages of diseased plants increased with increased spacing or with
decreased number of plants. A larger number of infected plants was
found, however, with the 15-inch spacing of 18,150 plants per acre than
with 12- or 18-inch spacings. A resolution for this apparent difference
between percentages of diseased plants and numbers of diseased plants
per acre is not attempted here.

Data on plant spacing of peanuts, receptors of the rosette virus disease inoculum vectored by an aphid, *Aphis craccivora* Koch., were given
by A'Brook (1964). Results were data based on "stand counts" per acre
and final numbers of diseased plants. Data were pooled from the years
1961, 1962, and 1963 and are given in table 202.

Table 202
Plant density and incidence of rosetting
(data from A'Brook, 1941)

Density of plants (no. per acre; 000 omitted)	9.6- 10.4	19.1- 21.4	36.0- 38.6	68.8- —	78.8- 106.6	160.2- 171.4	344.1- —
Density rosetted	1.0:1	1.1:1	2.1:1	3.5:1	5.5:1	10.3:1	40.1:1

Rosette diseased plants practically equalled the plant density at 10,000 plants per acre but decreased with plant density increase. At 344,100 plants per acre, the greatest density observed, healthy plants exceeded rosetted plants about 40:1. Although many factors affected inoculation of the plants with the virus, the author believed the incidence was attributable to the landing response of the aphids as related to the influence of ground cover.

Virus disease spread was less frequent where 600 pounds of bean seed were planted than where 50 pounds were used, according to Heathcote (1960). Data on leaf roll and mosaic are given in table 203.

Table 203
Rate of bean seeding and resulting leaf roll and mosaic incidences
(data from Heathcote, 1960)

Disease	Rate of seeding (lbs. /A.)	
	50	600
Leaf roll	7.7% (38.5)	0.3% (18)
Mosaic	6.5 (32.5)	1.1% (6)

Lower percentages and fewer infected plants (by simple multiplication computations, results in parenthesis) were present where more bean seeds were planted. Conjectures for this action may include, assuming plots were equal in size, (1) more movement of vectors in the 50-pound plots, (2) more attraction, or at least receptivity of more vectors by plants seeded, to 50 pounds, or (3) if plots received the same number of vectors, the plants in the 50-pound-per-acre rate were easier to inoculate or were more receptive to virus inoculation.

Hybridization of turnips was found affected by distance from contaminant pollen source and to row-mass numbers of receptor plants. Distance effects are shown elsewhere in data from Bateman (1947a). Such effects are emphasized in figure 45, in which more hybridization is shown at the 4.5 foot distance for each of the 1, 3, 6, and 12 rows than at the 15.0 foot distance. Although similar curvilinearities are shown for the two distance classifications, a flatter curve is seen for the 15.0 foot distance than for the 4.5 foot distance.

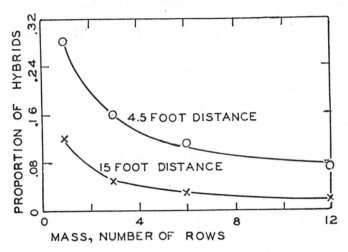

Fig. 45. Relationship of distances from a contaminating pollen source and to row masses of receptor plants.

Increasing numbers of receptor plants at each distance had decreasing amounts of hybridization. Such increase occurred at 4.5 and 15.0 feet from the contaminant source and at somewhat similar rates. There is close agreement of observed and curve values.

A tabulation is made of the hybridization data given by Bateman (1947a) from regressions computed by the author. Computed or expected numbers are given for hybridization in the different row masses at given distances from the contaminant source, with regression coefficients in table 204.

Table 204

Hybridization of turnips as manifold receptor at distances
from the contaminant, based on regression coefficients
(data from Bateman, 1947a)

Distances from contaminant (ft.)	Row mass receptors				Regression Coefficients
	1	3	6	12	
1.0	0.661	0.509	0.414	0.318	-0.317
2.5	0.389	0.267	0.193	0.117	-0.251
4.5	0.267	0.180	0.119	0.058	-0.202
15.0	0.110	0.065	0.037	0.009	-0.093
30.0	0.036	0.018	0.006	-0.006	-0.039

Multiplication of the receptor plants was related to contamination and in different rates, perhaps, to distance increases. Constant reductions in regression coefficients indicate the increasing flattening rates of curves by distance effects. In view of the close agreements of observed and curve values, considerable confidence may be placed in these observations.

Density of a pine bark beetle, *Ips confusus* (Lec.), in ponderosa pine, *Pinus ponderosa* Laws, influenced the number of beetles attracted to them, according to Gara (1963). Beetles were released upwind and downwind of logs infested with 8, 25, 37, 100, 250, 100, 37, 25, and 8 beetles, placed in order in a line at intervals of 8 m. Beetles passed logs with 8, 25, and 37 beetles on the downwind end and (perhaps because of a stronger attraction still downwind) also the second 100 beetle log attractant upwind from the 250-beetle lure. Based on averages, percentages of upwind and downwind attractants, the results are given in the following table.

Average number of beetles recovered
(data from Gara, 1963)

Beetles in log (no.)	250	100	37	25	8
Beetles recovered (%)	33.9	1.7	4.1	2.5	0

Effects of massed bettles is indicated in which 250 beetles attracted many more than 100, 37, or 25 beetles. Differences between the logs infested with 250 compared with 100, 37, or 25 beetles were probably significant.

In a laboratory study of a crayfish, *Cambarus alleni* Faxon, density in relation to dispersal was reported by Bovbjerg (1959). Results were based on the time spent in movement of 400 cm. following the release of different numbers of crayfish released. These are given in table 205.

Table 205
Movement time from densities of crayfish
(data from Bovbjerg, 1959)

Crayfish density (no.)	5	10	20
Time to move 400 cm. (min.)	20	15	10

Members in the greatest density moved most rapidly; those of least density, least rapidly. This suggested that crowding resulted in population shifts owing to suboptimal conditions.

Intraspecific density. Spatial relations are present with all organisms. Life functions are conducted in relationships of greater or lesser distances separating individuals. Relativity of space for organisms, therefore, is a recognized factor. Many species change patterns of dispersion with lapse of time, according to Watanabe *et al.* (1952). These were given as "dispersion," "grouping," and "nomadism." The term *isolation* is preferred to nomadism.

In the first pattern, dispersion refers to the exodus or emanation from origin or source. In grouping, discussed in chapter 7, the activity refers to the concentrating of individuals that often follows the initial exodus. Mating may occur during grouping. Isolation may be considered a disperse phase moving to favorable breeding or reproductively developmental sites. Organisms in the process of reproduction or actively preparing for reproduction may often require more space and, therefore, become more distantly spaced from other members of the race.

Effects of spacing of male-sterile onion bulbs at distances from pollen-fertile plants indicated very small differences in pollination between 6- and 18-inch spacings. These effects were found by Erickson and Gobleman (1956), who gave data from three separate experiments, which were used to calculate regression curves. Expected or smoothed curve data from these regressions appear more useful to show spacing effects. These data are given in table 206.

Table 206

Pollination of onions by spacing and in separate experiments
(data from Erickson and Gobleman, 1956)

Experiment	Spacing	Distance from pollen source								
		1	*4*	*7*	*10*	*13*	*16*	*19*	*22*	*25*
1952 A	6-inch	632	517	471	441	420	403	388	376	366
	18-inch	740	574	507	464	432	407	387	369	354
1952 B	6-inch	525	418	375	348	328	312	299	287	278
	18-inch	552	435	388	357	335	318	303	291	280
1952 B	6-inch	338	198	141	106	79	58	40	25	12
	18-inch	326	190	135	99	74	53	36	22	9

Rates of regression are similar, indicating equal pollination dispersal of onions in 6- and 18-inch spacings. Slightly more seed per head, however, were produced in the 18-inch spacing (467) than in the 6-inch spacing (454).

Crowding of houseflies, as determined by cage tests, was found by Morris (1964) to affect dispersion. Four strains of flies were used in each of four time units in dispersal movements from one cage to another. Cages were placed one beside the other and were interconnected with the one above with one-inch inside diameter polyethylene tubes of half-inch length, through which flies were given access to the other cage for predetermined intervals. Data given were pooled to indicate the effects of population density as per 100 flies changing cages per hour access time and are given in table 207.

Table 207
Cage populations of flies and dispersal time
(data from Morris, 1956)

Cage population	Hours access time			
	1	*2*	*4*	*6*
100	19.7	16.3	9.6	7.4
500	19.5	14.4	9.1	7.2
1,000	17.8	22.4	14.1	10.8

Significantly more flies passed from the 1,000-fly cage than from the 100-fly cage. Most flies in each population group changed cages the first or second hour. This suggests fatigue or lack of stimuli to move. More flies changed cages in cages of 1,000 population than in the cages of 100 and 500 flies, except for the one-hour access time period.

Population density of the rice plant skipper, *Parnara guttata* Bremer and Gray, was studied by Yosida (1954). He concluded that population density decided the pattern of distribution. He reported that the coefficient of diversity, s^2/x, increased gradually with increase of density or number of insects per plant. High population densities provided (1) higher mortality and (2) higher proportions of number of pupae to total number of insects.

Increasing the number of organisms of a species at a point, or in a small localized area, may result (1) in an increase in numbers of organisms surrounding the point, or (2) in extending the increase to areas more distant than ordinarily occurs, or (3) both. In studying foraging of honeybees, Levin (1961) found more cordovan bees per square yard at 110 (0.368) than at 330 yards (0.182) from a 20-colony apiary. After a 10-colony increase to 30 colonies, and increased number of bees per square yard (0.429) was found at the 110 yard distances, and fewer (0.113) at the 330 yard distance. Foraging in the nearby parts of the field was apparently uncrowded before the addition of the 10 colonies. An implication is evident that with only 20 colonies on the site, available forage was unused. If true, this implies that honeybees forage at 330 yards, regardless of crowded or uncrowded nearby forage plants.

Spatial relationships occur among organisms and must be recognized. Such relationships probably influence the distances to which individuals disperse.

It was observed by Kennedy *et al.* (1967) that individuals of *Drepanosiphum platanoides* (Schrank) were "patchily distributed" among whole leaves and on areas of single leaves. They tend to position themselves evenly, equidistantly from their neighbors, millimeters apart and in discrete groups. Data were given to show the nearest neighbor relationship (table 208).

Table 208
Drepanosiphum platanoides on sycamore leaves
classed according to distances from nearest neighbor
(data from Kennedy *et al.*, 1967)

Nearest neighbor (mm.)	2-3	4-5	6-7	8-9	10-13	14-28
Number of individuals	7	136	137	81	48	36

Most aphids settled 4-7 mm. apart. These data indicate spacing selection by the aphids for their living room.

Fly larvae of *Sarcophaga barbata* responded to spatial relationships with time of entrance into pupation, according to Zinforlin (1969), (table 209).

Table 209
Larvae in circular dishes of different diameter and days to pupation
(data from Zinforlin, 1969)

Dish diameter (cm.)	2.0	3.5	5.5	9.5	(Control)
Days to pupation (no.)	6.8	6.5	3.9	3.4	3.6

Most days were required for pupation at dish diameters of 2.0 and 3.5 cm., the smallest dishes. A doubling of days to pupation was found with a decrease from 9.5 to 2.0 cm. in diameter.

Density of population of the white pine cone weevil, *Conophthorus coniperda* (Schwarz), affects grouping considerably, as is shown by Henson (1961). Greater densities of weevils increased the formation of aggregations to 120 per dish, after which increases became very slight. Extreme variations in indexes of aggregations at the lowest number per dish (10 insects) suggested that accidental collisions are affected by distances between nearest neighbors. Through the use of dummies density was seen as a simple related effect, which is indicated by the data in table 210.

Table 210
Population densities affect grouping,
(data from Henson, 1961)

Density	Mean index*	Error
40 insects	0.387	0.122
80 insects	0.270	0.016
20 insects + 20 models	0.368	0.105
40 insects + 20 models	0.288	0.103

*An index of 0 is complete aggregation; an index of 1 is random association.

Weevils aggregate equally, or nearly so, with dummies as with insects. Pieces of wire, rice grains, and a variety of other objects (of about the same size and shape as the weevils), including other insects, living or dead, affect grouping. Such close proximity, contact or collision, is evidence of *thigmotropism*. Grouping of weevils ranged from immediate to incomplete aggregation in an hour, with most grouping occurring during the first few minutes. It was suspected that light stimulated movement and increased the likelihood of collision, since aggregation was greater in the dark than in a daylight illuminated room, Henson (1961).

Sex. Grouping may result from the presence of the opposite sex. Sexual differences may be the initial motivation for the action of grouping. Aggregation of males and females of the white pine cone weevil, *Conophthorus coniperda* (Schwarz), differed by degrees (Henson, 1961), as shown in table 211.

Table 211
Aggregation of sexes of the white pine cone weevil
(data from Henson, 1961)

Sex	Mean index*	Error
Male	0.346	0.078
Female	0.483	0.039

*An index of 0 is complete aggregation; an index of 1 is random association.

Movement of males may be more rapid than that of females. It was suspected that thigmotropism was more active in females during or following dispersal. Loss of or reduction of thigmotropism was suspected only following emergence from the overwintering cones to cones where broods were established. This phase of dispersal is present for a few days during which some stimulus apparently superseded thigmotropism.

Unlike most other insect species, honeybee colonies may be increased or decreased in an apiary to increase or decrease the populations in an area. This is in the highly specialized study of insect dynamics and needs further study. Comparison of a 5- and a 12-colony group of colonies was made by Hutson (1926) to indicate distance effects, with number of honeybees at distances with data in table 212.

Table 212
Dispersal of honeybees at distances from 5- to 12- colony groups
(data from Hutson, 1926)

Yds. from colony group	25	50	75	100
5-colony group	6.0	7.6	6.6	7.5
12-colony group	128.3	83.0	47.5	22.8

No response is seen in the 5-colony group to distance effects. More than five times as many bees were found at 25 as at 100 yards in the 12-colony group in response to distance. Observed and curve values were in close agreement in regression curves drawn of the 5-and 12-colony groups by Wolfenbarger (1946).

Centripetal movement. Movements of the tsetse fly, *Glossinia morsitans* Wst., were studied by Jackson (1940, 1941, and 1946). He reported that flies live longer in rainy than in dry seasons. Flies might be expected to live longer in habitats of shade and dampness, therefore, than in habitats of dryness. Habitats of dampness might also be expected to affect dispersal activities by retaining flies. Ambits of movement were believed restricted by "lines of vegetation types." Flies move in *ambits,* constantly going and coming, according to Jackson (1941). In the constant meandering, flies depart and some reenter haunts usual for the flies and terminate dispersion at a short distance, perhaps within one-half mile of the point of initiation of dispersion.

Reference was made by Jackson (1941) to an observed "centripetal tendency" of flies. This may mean that while the flies are motivated to move distantly there is also a motivation to move toward the development or origination site. Such motivation may arise from memory or from satisfactory states of life or of being in the habitat in which the organism developed. Insects other than tsetse flies may also exhibit centripetal movements. Such movement may be a characteristic of active dispersers and not of passive moving organisms. There may be some relationship of grouping or aggregation, on occasions, and centripetal force or tendency. In some or many instances there may be responses to the same or to similar stimuli. In these instances different terms may be used to express the same response. The term *grouping* might be the more expressive.

8

Internal Factors

Dispersion of organisms is influenced in part or principally by internal factors. These probably initiate dispersal. Internal factors include (1) aging, (2) nutrition, (3) periodicity, (4) sex or mating, (5) species (variety, kind or strain), and (6) stage (generation, phase or activity). Instincts (7), or "aggregation" and "imitation" instincts as discussed by Fraenkel (1932), are very likely the most important factors affecting the overall manifestations of all active disperser species.

Factors originating internally generally tend to manifest on all-or-none characteristic more than those originating externally. Evaluations of the various factors, however, have not been critically determined. More space is given below to dispersal that is considered to originate internally than to that which originates externally. Such instances are (1) more commonly reported, (2) perhaps of more importance, and (3) recognized by more students of dispersal than are externally motivated dispersion.

An "inherent" tendency to disperse was one of five factors given by Schoof and Siverly (1954) as affecting dispersion of the housefly. Reduction of vigor by inbreeding and restoration of vigor by crossing inbred lines are internal activations. Turner (1960) implied that these breeding techniques affected the dispersal distances of the larger milkweed butterfly, *Oncopeltes fasciatus* Dal.

Grouping. Grouping in response to internal activation is apparently similar in form, advantages, and achievements to that from external activation discussed above. Such grouping is frequently regularly periodical or rhythmical. Grouping action exhibited by living organisms probably has some relationship to the dispersion of organisms, according to Brown (1959), and is a form of internal activation. There was active aggregation on the part of *Aphis fabae* Scop. to form groups of aphids, which may be an aid in host selection, according to Ibbotson and Kennedy (1951).

AGING

Aging (senescence) is common to all species of organisms. Members of dispersing populations age as they move or are moved from the site of their natality to other locations. Many individuals perish en route to situations favorable to propagate their race. Mortality was one of two factors (dilution was the other) in decreasing organismal density with increase in distance from the origin, according to Kettle (1951) and Wadley (1957). Aging was defined simply by Strehler (1960) as an increased liability to die, or an increasing loss of vigor with increasing chronological age or with the passage of the life cycle. One result of increasing loss of vigor with increasing chronological age might be the dispersal to shorter and shorter distances with increasing age.

Although different species possess different life spans, all possess dispersal stages of limited time periods. Certain pollens, for example, have short lives and must perform their function within minutes or hours after dispersal. Jones and Newell (1946) calculated that pollen remained viable for 100 seconds, during which time it would travel 80 rods as dispersed by a nine-mile-per-hour wind. Certain viruses also have comparatively short lives within their vectors. Age is, therefore, a very important factor in the dispersal of viruses. Certain seeds, on the contrary, age slowly and may remain viable many years, presenting opportunities for widespread dispersal, awaiting favorable media for growth and reproduction. There are data showing dispersal effects of aging of active disperser organisms but none are known of organisms that disperse passively.

Extensive regression studies of a midge, *Culicoides impunctatus* Goet., populations and observed densities were given by Kettle (1951a). Mortality (attributable to age and to enemies) was given graphically also by Kettle (1951a) as its regression on distance from the breeding site (fig. 46). Mortality increased with distance increase (and age of dispersing adults) as a linear relationship between percentage mortality in terms of probits and logarithm of distance. If there were no mortality, reasoned Kettle (1951a), midges would move outward initially and then some would return to the breeding site, as shown in figure 47. Because of mortality, however, relative outward movements were greater than return movements. This increased the adult densities at distances away from the breeding site and reduced the size of the regression coefficient. This tends to increase organisms in the periphery of the dispersal range and to provide equalization of individuals over a wider range.

Aedes albopictus (Skuse) mosquitoes were indicated as dispersing fewer yards per day with increasing age, by Bonnet and Worcester (1946). From the data given on individual mosquito releases and recoveries a summary was prepared to show the average distance from release to recovery site. This is given in table 213.

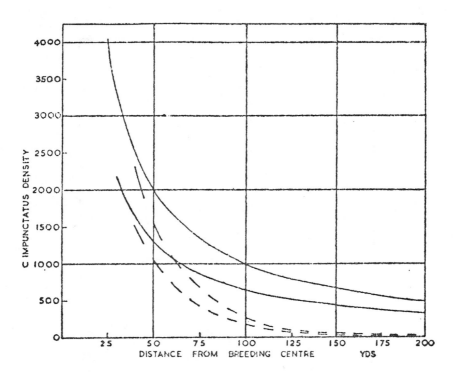

Fig. 46. Effect of "mortality" on the density of *Culicoides impunctatus* at distances from the breeding center. Actual density is indicated by the broken lines. If there were no mortality, midge density might be expected by the solid regression lines. Sex differences are indicated by the pairs of curves in which the upper curves refer to males and the lower ones to females (fig. 12 from Kettle, 1951a).

Table 213

Decelerating yards of dispersal with age of A. *albopictus*
(data from Bonnet and Worcester, 1946)

Minimum days between release and recapture	0	1	2	3	4	5	6	7	8	9	10	14	19	20	21
Number of yards mosquito per day	86	68	27	76	43	—	—	15	16	8	4	6	12	3	11

These data suggest a five-fold reduction in dispersal distance in a week and a ten-fold reduction in two weeks. The authors warned that "extreme caution" must be used in drawing conclusions from the small numbers of insects recovered and that these few were subject to influences

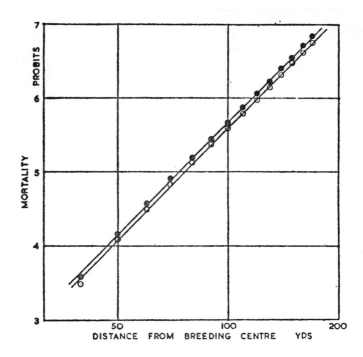

Fig. 47.   Mortality of *Culicoides impunctatus* as influenced by distance. The upper curve refers to male mortality, the lower to female mortality. The ordinate, mortality, is given in probits (fig. 13 from Kettle, 1951a).

such as wind, local situations, breeding and feeding sites, and many other factors. Equalization of organisms is recognized, however, and is discussed elsewhere (figs. 9, 10).

Rather extensive data were given on the dispersion of *Drosophila pseudoobscura* Frowola by Dobzhansky and Wright (1943). Day by day recovery data were given by which the effects of aging and dispersal rates were calculated. Estimated average distance of fruit fly dispersal from release to recovery points on successive days were used to indicate dispersal distances. Average cumulative distances on days 1, 2, 3, 4, and 5 were 90.0, 129.6, 141.8, 149.4, and 161.4 m., respectively. The relationships tend to show slightly less distance dispersed by each successive day's movements. The highest average dispersal occurred the first day. Curvature toward flatness (no dispersal) is indicated on third, fourth, and fifth days. Such flatness would appear to accelerate with the sixth and later days' collections and to terminate with death of the flies.

Data from a number of the experiments by Dobzhansky and Wright (1943) were arranged according to distance from the release site and day number following release. Regression studies were made from these data and are summarized as flies per trap-day in table 214.

Table 214
Meters distance traveled from release site
(data from Dobzhansky and Wright, 1943)

Day after realese	Distance from release site (meters)					
	10	50	90	130	170	210
1	65.1	13.1	5.3	3.7	2.5	1.6
2	33.5	15.2	5.1	3.6	2.4	1.5
3	21.5	9.5	4.9	3.4	2.4	1.5
4	14.8	6.3	4.7	3.3	2.3	1.5
5	10.4	3.9	4.6	3.2	2.2	1.4
6	7.1	6.0	4.4	3.0	2.1	1.4
7	4.6	3.3	4.2	2.9	2.1	1.4
8.5	1.7	1.6	3.9	2.7	2.0	1.3

Largest reductions over the nine-day period were at the 10-meter distance, and fewest reductions were at the 210-meter distance. Reductions occurred, however, at each distance class, with most (except at the 50-meter distance) recoveries on the first day at each distance class.

RHYTHMIC PERIODICITY

In the above discussion (Chapter 5) on "Light and Darkness" as an inorganic factor affecting dispersion, certain seeds and pollens were shown to disperse principally at certain hours. Maturity and "readiness" for dispersal were doubtless responsible for at least part of these movements. Temperature, light, and other factors may have coincided to bring about maturity and readiness. Initiation of dispersal (and of other activities) occurs at certain hours with considerable regularity. Although many external factors doubtless affect such dispersive movements, an internal factor that may be termed "rhythm" is of importance. It is usually impossible, or at least difficult, to determine whether external factors or rhythmic periodicity are the more responsible factors affecting dispersal. There is synchronization of individuals (aphids, as reported by Johnson and Taylor [1957b], for example), and they tend to disperse simultaneously. There appears to be no data, however, to show relationships of dispersal distances and periodicity rhythms.

Dispersal Distances of Small Organisms

Parasitization undoubtedly affects the ability of organisms to move or to be moved distances to which their dispersal usually occurs. Definitive data were given by Henderson (1955) on the relationship of parasitization and distance of dispersal. Percentage of parasitization of the female sugar beet leafhopper, *Circulifer tenellus* (Baker), was shown to decrease with distance traveled. The rate of such decrease is shown in figure 48, in which a straight-line relationship form of regression is

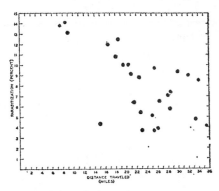

Fig. 48.   Parasitization of female beet leafhoppers in beetfields at distances traveled from the source, through spring movement (fig. 6 from Henderson).

suggested by the scatter. A closer alignment of observations is more desirable, however, for a more positive indication.

Parasitized individuals and gravid females of the sugar beet leafhoppers were found by Lawson *et al.* (1951) to fly nearer ground level than nonparasitized or nongravid females. Data in evidence are given in the following table.

Parasitization of the sugar beet leafhopper at elevations
(data from Lawson *et al.,* 1951)

	Height of leafhopper collection (feet)		
	2.5	*15.0*	*32.0*
Parasitized (percentage	3.5	1.9	0.7
Gravid females	7.4	8.1	4.6

The highest percentage of parasitized females was recorded at 2.5 feet elevation. It is suggested that such leafhoppers were flying at minimum heights to avoid obstructions. A higher percentage of gravid females was taken at 15.0 than at 2.5 or 32.0 feet elevations. This suggests that a peak of abundance is reached between 2.5 and 15.0 feet of elevation (and reached to lower percentages in excess of 32.0 feet).

Loss of life in dispersal journey. There are terrific losses of life during the dispersal journey, as is generally recognized by biologists.

Although inorganic and organic factors originating internally and externally are responsible for loss of life, they are placed in the above headings. Such losses may be envisaged by consideration of the immense amounts of corn pollen, for example, that are produced and the little that terminally falls on silks to consumate fertilization. Such losses are not known to have been numerically determined.

Life tables for man are generally recognized, but they are, for small organisms, rare or nonexistent. One life table determination, that for the spruce budworm, *Choristonura fumiferana* (Clem.), was given by Morris and Miller (1954), which recognized dispersion as an activity in which loss occurred. Data on losses attributable to dispersion were given under "Factor responsible" and "Dispersion, etc." for two age intervals, Instars I and II, for each of two plots in 1952-53. These are given in percentages (of organisms at the beginning of the instar) (table 215).

Table 215

**Life tables for the spruce budworm for instars I and II
(data from Morris and Miller, 1954)**

	Plot G4	*Plot G9*
Instar I	48	58
Instar II	63	70

Percentages of loss were higher in Plot G9 than in Plot G4 and for Instar II than for Instar I.

Sources of variation attributable to location are often very great, as is discussed in chapter 2 and as may be noted in many tables of analysis of variance that have been published. Loss of life of dispersing populations could also be expected to vary widely, depending on location. Losses of spruce budworm larvae were reported by Morris (1957) to vary according to location and ranged from 60 percent to 95 percent. Variations were attributed to host plant differences, density, and continuity of the stand. Moth dispersal losses were reported by Morris (1957) to vary according to location, foliage depletion, and weather.

That more organisms are lost in the disperse phases than in the more sedentary activities of a population is a generally accepted truism. This is one of the explanations for the "thinning out" of a dispersing population in the discussion by Wadley (1957) for the dispersion process. Hence, dispersion is intimately and closely related with the lives and deaths of organisms. Biological research of the population dynamics of the fruit-tree leafroller, *Archips argyrospilus* (Wlk.), by Paradis and LeRoux (1965) show amounts of reductions in disperse phases (table 216).

Table 216
**Percentage reductions of dispersing forms
of the fruit-tree leafroller
(data from Paradis and LeRoux, 1965)**

Year data were taken

Stage	1957	1958	1959	1960	1961	1962	1963	Mean
	Reduction of the stages indicated							
Instars I-II	70.0	57.6	70.1	56.5	70.9	57.6	31.3	59.0
Instars III-V	31.7	80.3	86.3	64.1	37.5	51.2	54.0	57.9
Adults	12.0	11.8	12.0	12.0	12.0	12.0	11.7	11.9
	Reduction of the generation							
Instars 1-II	58.3	47.1	60.5	53.2	61.4	50.7	27.7	51.3
Instars III-V	7.9	27.8	22.2	25.1	6.0	14.4	31.2	19.2
Adults	0.8	0.2	0.2	0.8	0.2	0.4	0.8	0.5

Rather similar percentages of reductions were observed for each stage of life each of the seven years. Larger percentage losses occurred in instars I-II than in the older stages. Larger losses to the generation (the species is univoltine in the area), however, were much less in instars III-V and still less in the adult stage than in instars I-II.

For the European corn borer, *Ostrinia nubilalis* (Hubn.), percentage losses, as shown by LeRoux *et al.* (1963) were 8.5 in instars I-II due to "dispersion," 23.4 in instars III-V due to "migration," 34.8 in instar V due to "migration," and 93.6 loss of adults due to "migration."

State of readiness. A state of maturity or readiness must exist in organisms before they initiate the dispersion journey. For dispersal, a species must have developed a preparation or ability for movement. Movement of many, or even most, species is impossible or is achieved with great difficulty or infrequency during certain life stages. Some species or kinds of bacteria and viruses may be ready, however, to initiate dispersal at practically all times.

Spore dispersal was discussed by Ingold (1939) and Dobbs (1942), pollen dispersal by Erdtman (1943), and larger plants in a more general reference by Guppy (1917). Light, temperature, and other weather factors were discussed by Johnson and Taylor (1957) in relation to aphid dispersal. State of readiness for dispersal by insects appears most generally discussed by authors for individual species. Practical application of readiness is experienced most frequently by quarantine inspectors, entomologists, plant pathologists, breeders, seedsmen, nurserymen, plant producers, greenhouse keepers, pet shops, and others dealing with populations of organisms.

References

Abbott, Charles H. 1950. Twenty-five years of migration of the painted lady butterfly, *Vanessa cardui*, in southern California. *Pan-Pacific Entomol.* 26(4):161-172.

A'Brook, John. 1964. The effect of planting date and spacing on the incidence of groundnut rosette disease and of the vector, *Aphis craccivora* Koch, at Mokwa, Northern Nigeria. *Ann. Appl. Biol.* 54:199-208.

Afzal, Mohammed and Abdul Hamid Khan. 1950. Natural crossing in cotton in Western Punjab. II. Natural crossing under field conditions. *Agronom. J.* 42(2):89-93.

Allee, W. C., A. E. Emerson, O. Park, T. Park and K. P. Schmidt. 1949. *Principles of animal ecology.* W. B. Saunders Co. Philadelphia. 837 p.

Andrewartha, H. G. and L. C. Birch, 1954. *The distribution and abundance of animals.* Univ. of Chicago Press. Chicago. 782 p.

————. 1961. *Introduction to the study of animal populations.* Univ. of Chicago Press. 281 p. Chicago.

Andrews, F. W. 1936. Investigations on black-arm disease of cotton under field conditions. I. The relation of the incidence and spread of black-arm disease of cotton to cultural conditions and rainfall in the Anglo-Egyptian Sudan. *Empire J. Exptl. Agric.* 4:344-356.

Angstrom, A. 1930. On the transportation of dust from low to high latitudes through the circulation of the atmosphere. *Geografiska Ann.* 12:88-92.

Annand, P. N. 1931. Beet leaf hoppers annual emigrations studied in desert breeding areas. *USDA Yrbk. 1931:* 114-116.

————, J. C. Chamberlain, C. F. Henderson and H. A. Waters, 1932. Movement of the beet leafhopper in 1930 in southern Idaho. *USDA Circ.* 244:1-24.

Anonymous. 1917. Mosquitoes and malaria. *Sci. Amer.* 117:174.

Ashby, Eric. 1948. Statistical ecology. II. A reassessment. *Botan. Rev.* 14:222-234.

Baas, J. 1959. The Mediterranean fruit fly, *Ceratitis capitata* Wied., in central Europe (Part 2). *Hofchen-Briefe* 12(3):113-140.

Bagnold, R. A. 1942. *The physics of windblown sand and desert dunes.* Methuen & Co., Ltd. London. 265 p.

Bailey, S. F., D. A. Eliason and B. L. Hoffman. 1965. Flight and dispersal of the mosquito *Cutlex taralis* Coquillet in the Sacramento valley of California. *Hilgardia* 37(3):73-113.

Bailey, Stanley F. and David C. Baerg. 1967. The flight habits of *Anopheles freeborni* Aitkin. *Calif. Mosquito Control Assn. Proc.* Paper No. 35:55-69.

Baker, R. R. 1968. Sun orientation during migration of some British butterflies. *Proc. Roy. Entomol. Soc. London.* Ser. A. *General Entomol.* 43(7/9):8995.

Baker W. A., W. G. Bradley and C. A. Clark. 1949. Biological control of the European corn borer in the United States. *USDA Tech. Bull.* 983-1-185.

Barber, George W. and Eleanor B. Starnes. 1949. The activities of house flies. *J. New York Entomol. Soc.* 57(4):203-214.

Basden, E. B. and D. G. Harnden. 1956. Drosophilidae (Diptera) within the Arctic Circle. II. The Edinburgh University Expedition to Subarctic Norway, 1953. *Roy. Entomol. Soc. London* 108:147-162.

Bassett, John. 1959. Surveys of air-borne ragweed pollen in Canada with particular reference to sites in Ontario. *Canadian J. Plant Sci.* 39:491-497.

Bateman, A. J. 1947. Contamination of seed crops. III. Relation with isolation distance. *Heredity* 1:303-336.

————. 1950. Is gene dispersion normal? *Heredity* 4:353-363.

Bawden, F. C. 1951. Plant Pathology Department. In: *Rothamstead Experimental Station Report for 1950:* 69-78.

Beall, Geoffrey and Richard R. Rescia. 1953. A generalization of Neymans contagious distributions. *Biometrics* 9(3):354-386.

Bedard, W. D. 1939. Mountain pine beetle infestation possibly influenced by road construction. *Bureau Entomol. and Plt. Quarantine News Letter* 6(5):8.

Berland, Lucien. 1935. Premier resultats des mes recherches en avion sur la faune et la flore atmospheriques. *Ann. Soc. Entomol. France* 104:73-96.

Bishopp, F. C. and E. W. Laake. 1921. Dispersion of flies by flight. *J. Agric. Res.* 21:729-766.

Bitancourt, A. A. and H. S. Fawcett. 1944. Statistical studies of distribution of psorosis-infected trees in citrus orchards. *Phytopathology* 34:358-375.

Blair, I. D. 1943. Behaviour of the fungus *Rhizoctonia solani* Kühn in the soil. *Ann. Appl. Biol.* 30:118-127.

Blake, George H., Jr. 1958. The influence of honey bees on the production of crimson clover seed. *J. Econ. Entomol.* 51:523-527.

Blencowe, J. W. and T. W. Tinsley. 1951. The influence of density of plant population on the incidence of yellows in sugar-beet crops. *Ann. Appl. Biol.* 38:395-401.

Bonnet, David D., and Douglas J. Worcester. 1946. The dispersal of *Aedes albopictus* in the Territory of Hawaii. *Amer. J. Trop. Med.* 26(4):465-476.

Bosanquet, C. H., and J. L. Pearson. 1936. The spread of smoke and gases from chimneys. *Trans. Faraday Soc.* 32:1249-1264.

Bovbjerg, Richard V. 1952. Ecological aspects of dispersal of the snail *Campeloma decisum* Say. *Ecol.* 33:169-176.

––––––. 1959. Density and dispersal in laboratory crayfish populations. *Ecol.* 40:504-506.

––––––. 1960. Behavioral ecology of the crab, *Pachygrapsus crassipes.* *Ecol.* 41:785-790.

Boyce, Stephen G. 1954. The salt spray community. *Ecology Monograph* 24:29-67.

Braun, E., R. M. MacVicar, D. R. Gibson, P. Pankiw and J. Guppy. 1953. Studies in red clover seed production. II. *Canadian J. Sci.* 33:437-447.

Briggs, John C. 1967. Dispersal of tropical marine shore animals: Coriolis parameters or competition. *Nature* 216:350.

Broadbent, L., and P. H. Gregory. 1948. Experiments on the spread of rugose mosaic and leaf roll in potato crops in 1946. *Ann. Appl. Biol.* 35(3):395-405.

––––––. 1949. Factors affecting the activity of alatae of the aphids *Myzus persicae* (Sulzer) and *Brevicoryne brassicae* (L.). *Ann. Appl. Biol.* 36:40-62.

Bromfield, K. R., J. F. Underwood, C. E. Peet, E. H. Grissinger and C. H. Kingsolver R. 1959. Epidemiology of stem rust of wheat: IV. The use of rods as spore collecting devices in a study on the dissemination of stem rust of wheat uredospores. *USDA Plt. Disease Reptr.* 43(11):855-862.

Brown, E. S. 1951. The relation between migration-rate and type of habitat in aquatic insects, with special reference to certain species of Corixidae. *Zool. Soc. London Proc.* 121:539-545.

––––––. 1962. Researches on the ecology and biology of *Eurygaster integriceps Put.* (Hemiptera, Scutelleridae) in middle East countries with special reference to the overwintering period. *Bull. Entomol. Res.* 53:445-514.

––––––. 1965. Notes on the migration and direction of flight of *Eury-*

gaster and Aelia species (Hemiptera, Pentatomidae) and their possible bearing on invasions of cereal crops. *J. Animal Ecol.* 34:93-107.

Brown, Frank A., Jr., 1959. The rhythmic nature of animals and plants. *Amer. Sci.* 47:147-168.

Buchanan, T. S. and J. W. Kimmey. 1938. Initial tests of the distance of spread to and intensity of infection on *Pinus monticola* by *Cronartium ribicola* from *Ribes lacustre* and *R. viscosissimum*. *J. Agric Res.* 56:9-30.

Buller, A. H. Reginald. 1924. *Researches on fungi.* Vol. III. Longmans, Green and Company. London. 611 p.

Burla, H., A. Brita da Cunha, A. G. L. Cavalcanti, Th. Dobzhansky and C. Pavan. 1950. Population density and dispersal rates in Brazilian *Drosophila willistoni. Ecol.* 31(3):393-404.

Butler, E. J. 1917. The dissemination of parasitic fungi and international legislation. *Mem. Dept. Agric. India* (Botan. Ser.) 9:1-73.

Calnaido, D., R. A. French and L. R. Taylor, 1965. Low altitude flight of *Oscinella frit* L. (Diptera: Chloropidae). *J. Animal Ecol.* 34:45-61.

Cammack, R. H. 1958. Factors affecting infection gradients from a point source of *Puccinia polysora* in a plot of *Zea mays. Ann. Appl. Biol.* 46:186-197.

Campos, E. G., H. A. Trevino and L. G. Strom. 1961. The dispersal of mosquitoes by railroad trains involved in international traffic. *Mosquito News* 21(3):190-192.

Carnegie, A. J. M. 1957. Intensities of scale infestation in citrus orchards under different conditions of cultivation. *Entomol. Mo. Magazine* 93:100.

Carter, Walter. 1929. The purpose of predicting outbreaks of *Eutettix tenellus* Baker under present-day conditions. *J. Econ. Entomol.* 22:154-158.

Chamberlin, T. R. 1941. The wheat jointworm in Oregon with special reference to its dispersion, injury and parasitization. *USDA Tech. Bull.* 784:1-47.

Chepil, W. S. 1958. Soil conditions that influence wind erosion. *USDA Tech. Bull.* 1185:1-40.

Christie, J. R. 1959. *Plant nematodes: Their bionomics and control.* Fla. Agric. Expt. Sta. Gainesville. 256 p.

Clark, Austin H. 1931. Some observations on butterfly migrations. *Sci. Mo.* 32:150-155.

Clark, D. P. 1962. An analysis of dispersal and movement in *Phaulacridium vittatum* (Sjost) (Acrididae). *Australian J. Zool.* 10:382-399.

Clark, Philip J. and Francis C. Evans. 1955. On some aspects of spatial pattern in biological populations. *Science* 121(3142):397-398.

————. 1956. Grouping in spatial distributions. *Science* 123(3192): 373-374.

Clarke, George L. 1954. *Elements of ecology.* John Wiley & Sons, Inc. New York 534 p.

Clements, F. E. and V. E. Shelford. 1939. *Bio-ecology.* John Wiley & Sons, Inc., New York. 425 p.

Clifford, H. T. 1959. Seed dispersal by motor vehicles. *J. Ecol.* 47:311-315.

Coad, B. R. and T. F. McGhee. 1917. Collection of weevils and infested squares as a means of control of the cotton-boll weevil in the Mississippi delta. *USDA Bull.* 564:1-51.

Collins, C. W. and W. L. Baker. 1934. Exploring the upper air for wind-borne gypsy moth larvae. *J. Econ. Entomol.* 27:320-327.

Common, I. F. B. 1954. A study of the ecology of the adult bogong moth, *Agrotis infusa* (Boisd.) (Lepidoptera: Noctuidae), with special reference to its behavior during migration and aestivation. *Aust. J. Zool.* 2(2):222-263.

Coons, B. F. and J. O. Pepper. 1968. Alate aphids captured in air traps arranged at different heights. *J. Econ. Entomol.* 61(5):1473-1474.

Coons, G. H., Dewey Stewart, H. W. Bockstahler and C. L. Schneider. 1958. Incidence of savoy in relation to the variety of sugar beets and to the proximity of wintering habitat of the vector, *Piesma cinerea. Plant Dis. Reptr.* 42(4):502-511.

Corrsin, S. 1961. Turbulent flow. *Amer. Sci.* 49:300-325. Reprinted from *Johns Hopkins Magazine.*

Coulter, J. M., C. R. Barnes and H. C. Cowles. 1931. *A textbook of botany.* Vol. 2. Ecology. American Book Company. New York. 499 p.

Cragg, J. B. and J. Hobart. 1955. A study of a field population of the blowflies *Lucilia caesar* (L.) and *L. sericata* (MG). *Ann. Appl. Biol.* 43(4):645-663.

Craigie, J. H. 1945. Epidemiology of stem rust in western Canada. *Sci. Agric.* 25:285-401.

Crane, H. L., C. A. Reed, and M. N. Wood. 1937. Nut breeding. *USDA Yrbk. of Agric.*:827-889.

Crossman, S. S. 1917. Some methods of colonizing imported parasites and determining their increase and spread. *J. Econ. Entomol* 10:177-188.

Daines, R. H., I. A. Leone and E. Brennan. 1960. Air pollution as it affects agriculture in New Jersey. *N.J. Agric. Expt. Sta. Bull.* 794:1-14.

Dalmat, Herbert T. 1950. Studies on the flight range of certain Simuliidae, with use of aniline dye marker. *Ann. Entomol. Soc. Amer.* 43: 537-545.

Darlington, Philip J., 1957. *Zoogeography: The geographical distribution of animals.* John Wiley & Sons. New York. 675 p.

Davies, W. M. 1936. Studies on the aphides infesting the potato crop. Laboratory experiments on the effect of wind velocity on the flight of *Myzus persicae* Sulz. *Ann. Appl. Biol.* 23:401-408.

Deane, L. M., R. G. Damasceno and R. Arouck. 1953. Distribuição vertical de mosquitos em uma floresta dos arredores de Belém. *Para. Folia Clin. et Biol.* 20(2):101-107.

DeLong, D. M. and O. S. Caldwell. 1935. Hibernation studies of the potato leafhopper (*Empoasca fabae* Harris) and related species of *Empoasca* occurring in Ohio. *J. Econ. Entomol.* 28:442-444.

Demski, James W. and John S. Boyle. 1968. Spread of necrotic ringspot virus in a sour cherry orchard. *Plant Disease Reptr.* 52:972-974.

Dice, Lee R. 1952. Measure of the sapcing between individuals within a population. *Contrib. Lab. Vert. Biol. Univ. Mich.* 55:1-23.

Dicke, R. H. 1959. Gravitation—an enigma. *Amer. Sci.* 47(1):25-40.

Dickson, R. C., Edward F. Laird, Jr. and George R. Pesho. 1955. The spotted alfalfa aphid (Yellow clover aphid) on alfalfa. *Hilgardia* 24(5):93-118.

Dobbs, C. G. 1942. On the primary dispersal and isolation of fungal spores. *New Phytol.* 41:63-69.

Dobzhansky, Th. and Sewall Wright. 1943. Genetics of natural populations. X. Dispersion rates of *Drosophila pseudoobscura. Genetics.* 28:304-340.

———— and ————. 1947. Genetics of natural populations. XV. Rate of diffusion of a mutant gene through a population of *Drosophila pseudoobscura. Genetics* 32:303-324.

———— and C. Pavan. 1950. Local and seasonal variations in relative frequencies of species of Drosophila in Brazil. *J. Animal Ecol.* 19(1): 1-14.

Dominick, C. B. 1943. Life history of the tobacco flea beetle. *Va. Agric. Expt. Sta. Bull.* 355:1-39.

Doolittle, S. P. and M. N. Walker. 1925. Further studies on the overwintering and dissemination of cucurbit mosaic. *J. Agric. Res.* 31:1-58.

Doucette, Charles F., Randall Latta, Charles Martin, Ralph Schopp and Paul M. Eide. 1942. Biology of the narcissus bulb fly in the Pacific Northwest. *USDA Tech. Bull.* 809:1-67.

Drake, D. C. and R. K. Chapman, Pt. I, L. N. Chiykowski and R. K. Chapman, Pt. II. 1965. Migration of the six-spotted leafhopper, *Macrosteles fascifrons* (Stahl). *Wisc. Agric. Expt. Sta. Res. Bull.* 261:1-45.

Durham, O. C. 1951. The pollen harvest. *Econ. Bot.* 5(3):211-254.

Eckert, J. E. 1933. Flight range of the honeybee. *J. Agric. Res.* 47:257-285.

Ederer, Grace Mary. 1965. Dissemination of bacteria by laboratory personnel. *Amer. J. Med. Tech.* Mar.-Apr.:108-120.

Eggler, Willis A. 1959. Manner of invasion of volcano deposits by plants, with further evidence from Paricutin and Jorulo. *Ecol. Monographs* 29(3):267-284.

————. 1963. Plant life of Paricutin volcano, Mexico, eight years after activity ceased. *Amer. Midland Natur.* 69:38-68.

Elton, Charles S. 1949. Population interspersion. An essay on animal community patterns. *J. Ecol.* 37(1):1-23.

————. 1958. *The ecology of invasions by animals and plants.* Methuen & Company. London. 181 p.

Epling, Carl. and Th. Dobzhansky. 1942. Genetics of natural populations. VI. Micrographic races in *Linanthus parryae. Genetics.* 27:317-332.

Erdtman, G. 1943. An introduction to pollen analysis. *Chronica Botanica.* Waltham, Mass. 239 p.

Erickson, H. T. and W. H. Gabelman. 1956. The effect of distance and direction on cross-pollination in onions. *Proc. Amer. Soc. Hort. Sci.* 68:351-357.

Etter, Alfred G. 1948. Seeds that ride livestock. *Missouri Bot. Garden Bull.* 36(10):170-172.

Evans, F. G. C. 1951. An analysis of the behaviour of *Lepidochitona cinereus* in response to certain physical features of the environment. *J. Animal Ecol.* 20(1):1-10.

Ewert, M. A. and H. C. Chiang. 1966. Dispersal of three species of coccinellids in corn fields. *Canadian Entomol.* 98:999-1003.

Faulwetter, R. C. 1917. Dissemination of the angular leafspot of cotton. *J. Agric. Res.* 8:457-475.

————. 1917a. Wind-blown rain, a factor in disease dissemination. *J. Agric. Res.* 10:639-648.

Fell, H. Barraclough. 1967. Resolution of coriolis parameters for former epochs. *Nature* 214:1192-1198.

Felt, E. Porter. 1937. Balloons as indicators of insect drift and of Dutch elm disease spread. *Bartlett Tree Res. Labs. Bull.* 2:3-10.

Fenton, F. A. and E. W. Dunnam. 1928. Dispersal of the cotton boll weevil, *Anthonomous grandis* Boh. *J. Agric. Res.* 36:135-149.

Ficht, G. A. and T. E. Heinton. 1939. Studies on the flight of European corn borer moths to light traps: A progress report. *J. Econ. Entomol.* 32:520-526.

————, ———— and J. M. Fore. 1940. The use of electric light traps in the control of the European corn borer. *Agric. Engin.* 21:87-89.

Fisher, Katherine. (Mrs. K. Grant) 1938. Migration of the silver-Y moth *(Plusia gamma)* in Great Britain. *J. Animal Ecol.* 7(2):230-247.

Fleschner, C. A. 1958. The effect of orchard dust on the biological control of avocado pests. *Yrbk. Calif. Avocado Soc.* 42:94-98.

Fletcher, T. Bainbridge. 1925. Migration as a factor in pest outbreaks. *Bull. Entomol. Res.* 16:177-181.

Foister, C. E. 1935. The relation of weather to fungus and bacterial disease outbreaks. *J. Econ. Entomol.* 29:923-940.

————. 1946. The relation of weather to fungus diseases of plants. II. *Bot. Rev.* 12(1):548-591.

Fracker, S. B. 1936. Progressive intensification of uncontrolled plant disease outbreaks. *J. Econ. Entomol.* 29:923-940.

Fraenkel, Gottfried. 1932. *Die Wanderungen der Insekten.* Julius Apringer. Berlin. 238 p.

Frampton, Vernon L., M. B. Linn and E. D. Hansing. 1942. The spread of virus diseases of the yellows type under field conditions. *Phytopathology* 32:799-806.

Freeman, J. A. 1938. Composition of the aerial/insect fauna up to 300 feet. *Nature* 142:153.

————. 1945. Studies in the distribution of insects by aerial currents. *J. Animal Ecol.* 14(2):128-154.

Frisch, Karl von. 1950. *Bees, their vision, chemical senses and language.* Cornell University Press. Ithaca. 119 p.

Fye, R. E., W. W. McMillan, R. W. Walker and A. R. Hopkins. 1959. The distance into woods along a cotton field at which the boll weevil hibernates. *J. Econ. Entomol.* 52: 310-312.

Gaines, J. C. and K. P. Ewing. 1938. The relation of wind currents, as indicated by balloon drifts, to cotton flea hopper dispersal. *J. Econ. Entomol.* 31:674-677.

Gara, R. I. 1963. Studies on the flight behavior of the *Ips confusus* (Lec.) (Coleoptera: Scolytidae) in response to attractive material. *Contr. Boyce Thompson Inst.* 22:51-66.

Gardner, Max William. 1918. The mode of dissemination of fungus and bacterial diseases of plants. *Ann. Rept. Mich. Acad. Sci.* 20:355-423.

Gardner, T. R. 1938. Influence of feeding habits of *Tiphia vernalis* on the parasitization of the Japanese beetle. *J. Econ. Entomol.* 31:204-207.

Geiger, Rudolf. 1950. *The climate near the ground.* Harvard Univ. Press. Cambridge. 482 p. (Transl. Milroy N. Brooks Stewart and others.)

Gillies, M. T. 1961. Studies on the dispersion and survival of *Anopheles gambiae* Giles in East Africa, by means of marking and release experiments. *Bull. Ent. Res.* 52(1):99-127.

Gilmour, Darcy, D. F. Waterhouse and G. A. McIntyre. 1946. An account of experiments to determine the natural population density of sheep blowfly, *Lucilia cuprina* Wied. *Commis. Aus. Council Sci. and Ind. Res. Bull.* 195:1-39.

Glass, Lynn W. and Richard V. Bovbjerg. 1959. Density and dispersion in laboratory populations of caddisfly larvae (Cheumatopsyche: Hydropsychidae). *Ecology* 50:1082-1084.

Glick, P. A. 1939. The distribution of insects, spiders, and mites in the air. *USDA Tech. Bull.* 673:1-150.

Glick, Perry A. 1967. Aerial dispersal of the pink bollworm in the United States and Mexico. *USDA Production Res. Rept.* No. 96:1-12.

Godwin, H. 1934. Pollen analysis. An outline of the problems and potentialities of the method. *New Phytol.* 33:278-305.

Goodall, D. W. 1952. Quantitative aspects of plant distribution. *Biol. Rev.* 27:194-245.

Goodwin, Melvin H. Jr. 1942. Studies on artificial resting places of *Anopheles quadrimaculatus* Say. *J. Natl. Malaria Soc.* 1:93-99.

Greenbank, D. O. 1956. The role of climate and dispersal in the initiation of outbreaks of the spruce budworms (*Choristoneura fumiferena*) in New Brunswick. *Canadian J. Zool.* 34:453-476.

————. 1963. The analysis of moth survival and dispersal in the unsprayed area. In *Mem. Entomol. Soc. Canada* 31:87-99.

Greenslade, R. M. 1941. The migration of the strawberry aphis *Capitophorus fragariae* Theob. *J. Pomology and Hort. Sci.* (London) 19:87-106.

Gregory, P. H. 1945. The dispersion of air-borne spores. *Trans. Brit. Mycol. Soc.* 28:26-72.

————. 1950. Deposition of air-borne particles on trap surfaces. *Nature* (London) 166:487.

————. 1951. Deposition of air-borne *Lycopodium* spores on cylinders. *Ann. Appl. Biol.* 38(2):357-376.

————, E. J. Guthrie and Maureen E. Bunce. 1959. Experiments on splash dispersal of fungus spores. *J. Gen. Microbiol.* 20:328-354.

————. 1961. *The microbiology of the atmosphere.* Interscience Publishers. New York. xv + 291 p.

————. 1962. Outdoor aerobiology. Smithsonian Institution. Rept. for 1961:445-453.

Greig-Smith, P. 1957. *Qualitative plant ecology.* Academic Press, Inc. London. 198 p.

Gressitt, J. Linsley. 1961. Problems in the zoogeography of Pacific and

Anarctic Insects. *Pacific Insects Monograph* 2:1-94. Bernice P. Bishop Museum. Honolulu.

Guppy, H. B. 1917. *Plants, seeds and currents in the West Indies and Azores.* Williams and Norgate. London. 531 p.

Hagen-Smit, A. J. 1958. Air conservation. *Science* 128:869-878.

Haine, Else. 1955. Aphid take-off in controlled wind speeds. *Nature* 175(4454):474-475.

Hampton, R. O. 1967. Natural spread of virus infections to beans. *Phytopathology* 57:476-481.

Hansing, E. D. and Vernon L. Frampton. 1940. The dissemination of yellow dwarf of potatoes and its leaf hopper vector, *Aceratagallia sanguinolenta. Phytopathology* 30:7.

Hardy, A. C. and P. S. Milne 1938. Aerial drift of insects. *Nature* 141: 602-603.

————— and —————— 1938a. Studies in the distribution of insects by aerial currents. *J. Animal Ecol.* 7(2):199-229.

Hartstack, Albert W., J. H. Hollingsworth and D. A. Lindquist. 1968. A technique for measuring trapping efficiency of electric insect traps. *J. Econ. Entomol.* 61:546-552.

Haskell, G. and P. Dow. 1951. Studies with sweet corn. V. Seed-settings with distance from pollen source. *Empire J. Exptl. Agric.* 19:45-50.

Heald, F. D., M. W. Gardner and R. A. Studhalter. 1915. Air and wind dissemination of the ascospores of the chestnut blight fungus. *J. Agric. Res.* 3:493-526.

Heape, Walter. 1931. *Emigration, migration and nomadism.* Heffer, Cambridge, England. 369 p.

Heathcote, G. D. 1960. Field beans. Rothamsted Expt. Sta. Report for 1959:97.

Henderson, Charles F. 1955. Parasitization of the beet leafhopper in relation to its dissemination in southern Idaho. *USDA Circ.* 968: 1-16.

Henson, W. R. 1959. Some effects of secondary dispersive processes on distribution. *Amer. Natur.* 43:315-320.

—————. 1961. Laboratory studies on the adult behavior of *Conopthorus coniperda* (Schwarz) (Coleoptera: Scolytidae). II. Thigmotropic aggregation. *Ann. Entomol. Soc. Amer.* 54:810-819.

Hervey, G. E. R. and C. E. Palm. 1935. A preliminary report on the response of the European corn borer to light. *J. Econ. Entomol.* 28:670-675.

Hirst, J. M., O. J. Stedman and G. W. Hirst. 1967. Long-distance spore transport: Vertical sections of spore clouds over the sea. *J. Genl. Microbiol.* 48:357-377.

Hodek, Ivo. 1960. Hibernation-bionomics in Coccinellidae. *Casopis Ceskoslovenske Spolecnosti Entomol.* 57(1):1-20.

Hofmann, J. V. 1911. Natural reproduction from seed stored in the forest floor. *J. Agric. Res.* 11:1-26.

Holbrook, R. F. Morton Beroza and Emory D. Burgess. 1960. Gipsy moth *(Porthetria dispar)* detection with the natural female sex lure. *J. Econ. Entomol.* 53:751-756.

Horsfall, J. G. and A. E. Dimond. 1960. Plant Pathology. III. *The diseased population: epidemics and control.* Academic Press. New York. 675 p.

Horsfall, William R. 1942. Biology and control of mosquitoes in the rice area. *Ark. Agric. Expt. Sta. Bull.* 427:1-46.

Hubert, Kurt. 1932. Beobachtung über die Verbreitung des Gelbrostes bei Künstlichen Feldinfektionen. *Fortschr. Landw.* 7:195-198.

Huffaker, Carl B. and Richard C. Back. 1945. A study of the effective flight range, density, and seasonal fluctuations in the abundance of *Anopheles quadrimaculatus* Say. in Delaware. *Amer. J. Trop. Med.* 25:155-161.

Hughes, John H. and John E. Porter. 1956. Dispersal of mosquitoes through transportation, with particular reference to immature stages. *Mosquito News* 16(2):106-111.

————. 1961. Mosquito interceptions and related problems in aerial traffic arriving in the United States. *Mosquito News* 21:93-100.

Hunter, W. D. 1910. The status of the cotton boll weevil in 1909. *USDA Circ.* Bureau of Entomology: 122:1-12.

————. and B. R. Coad 1923. The boll weevil problem. *USDA Farmers Bull.* 1329:1-30.

Hussey, N. W. and W. J. Parr. 1963. Dispersal of the greenhouse red spider, *Tetranychus uritcae* Koch (Acarina Tetranychidae). *Entomol. Exptl. and Appl.* 6:207-214.

Hutson, Ray. 1926. Relationship of the honeybee to fruit pollination in New Jersey. A preliminary report. *N. J. Agric. Expt. Sta. Bull.* 434:1-32.

Ibbotson, Alan and J. S. Kennedy. 1951. Aggregation in *Aphis fabae* Scop. *Ann. Appl. Biol.* 38:65-78.

Ingold, C. T. 1939. *Spore discharge in land plants.* Clarendon Press. Oxford. 178 p.

———— 1956. The spore deposit of *Daldinia*. *Brit. Mycol. Soc. Trans.* 39:378-380.

———— and Susan A. Hadland. 1958. The ballistics of Sordaria. *New Phytol.* 58(1):46-57.

Ishii, H. and H. Koyama. 1952. Conidial discharge and the secondary infection in the ear blight of cereals. (In Japanese.) (*Biol Abstr.* 27. Item 2390. 1952.)

Itô, Yosiaki. 1960. Ecological studies on population increase and habitat segregation among barley plants. *Natl. Inst. Agric. Sci.* (Tokyo) Ser. C (11): 45-130.

———— and Kozuyoshi Miyashita. 1961. Studies on the dispersal of leaf- and plant-hoppers. I. Dispersal of *Nephotettix cincticeps* Uhler on paddy fields at the flowering stage. *Jap. J. Ecol.* 11(5):181-186.

Jachowski, L. A., Jr. 1954. Filariasis in American Samoa. V. Bionomics of the principal vector, *Aedes polynesiensis* Marks. *Amer. J. Hyg.* 60(2):186-203.

Jackson, C. H. N. 1940. The analysis of a tsetse fly population. *Ann. Eugenics.* 10:332-369.

————. 1941. The economy of a tsetse population. *Bull. Entomol. Res.* 32:53-55.

————. 1946. An artifically isolated generation of tsetse flies (Diptera). *Bull. Entomol. Res.* 37(2):291-299.

Janssen, C. R. 1966. Recent pollen spectra from the deciduous and coniferous-deciduous forests of northeastern Minnesota: A study in pollen dispersal. *Ecology* 47:804-825.

Jenkins, D. W. and C. C. Hassett. 1951. Dispersal and flight range of subarctic mosquitoes marked with radiophosphorus. *Canad. J. Zool.* 29:178-187.

Jensen, L. and H. Bøgh. 1941. On conditions influencing the danger of crossing in the case of wind-pollinating cultivated plants. *Tidsskr. Planteaul.* 46(2):238-266.

Johnson, C. G. 1950. Infestation of a bean field by *Aphis fabae* Scop. in relation to the wind direction. *Ann. Appl. Biol.* 37(3):441-450.

————. 1951. The study of wind-borne insect populations in relation to terrestrial ecology, flight periodicity and the estimation of aerial populations. *Sci. Progr.* 39(153):41-62.

———— and H. L. Penman. 1951. Relationship of aphid density to altitude. *Nature* 168:337-340.

————. 1954. Aphid migration in relation to weather. *Biol. Rev.* 29:87-118.

————. 1957. The distribution of insects in the air and the empirical relation of density to height. *J. Animal Ecol.* 26:479-494.

———— and L. R. Taylor. 1957. Periodism and energy consummation with special reference to flight rhythms in aphids. *J. Exptl. Biol.* 34 (2):209-221.

————, ———— and T. R. E. Southwood. 1962. High altitude migra-

References
213

tion of *Oscinella frit* L. (Diptera: Chloropidae). *J. Animal Ecol.* 31(2):373-383.

———. 1969. *Migration and dispersal of insects by flight.* Methuen & Co. Ltd. London. 763 pp. + xxii.

Jones, Melvin D. and L. C. Newell. 1946. Pollination cycles and pollen dispersal in relation to grass improvement. *Nebr. Agric. Expt. Sta. Bull.* 148:1-42.

——— and James S. Brooks. 1950. Effectiveness of distance and border rows in preventing outcrossing in corn. *Okla. Agric. Expt. Sta. Tech. Bull.* T-38:1-18.

Joyce, C. R. 1961. Potentialities for accidental establishment of exotic mosquitoes in Hawaii. *Proc. Hawaiian Entomol. Soc.* 17(3):403-413.

Joyce, R. J. V. 1956. Insect mobility and design of field experiments. *Nature* (London) 1772(4502):282-283.

Keitt, G. W. and D. H. Palmiter. 1937. Potentialities of eradicant fungicides for combatting apple scab and some other plant diseases. *J. Agric. Res.* 55:397-437.

———, C. N. Clayton and M. H. Langford. 1941. Experiments with eradicant fungicides for combatting apple scab. *Phytopathology* 31:296-322.

Kennedy, J. S., Lorna Crawley and A. D. McLaren. 1967. Spaced-out gregariousness in sycamore aphids *Drepanosiphum platanoides* (Schrank) (Hemiptera: Callaphidae). *J. Animal Ecol.* 36(1):147-170.

Kettle, D. S. 1951. Some factors affecting the population density and flight range of insects. *Proc. Roy. Soc. of London* 26:59-63.

———. 1951a. The spatial distribution of *Culicoides impunctatus* Goet. under woodland and moorland conditions and its flight range through woodland. *Bull. Entomol. Res.* 42:239-291.

Kimball, Herbert H. and Irving F. Hand. 1924. Investigation of the dust content of the atmosphere. *Mo. Weather Rev.* 52:133-139.

Kligler, I. J. 1924. Flight of Anopheles mosquitoes. *Roy. Soc. Trop. Med. & Hyg.* 18:199-202.

Knipling, E. F. 1955. Possibilities of insect control or eradication through the use of sexually sterile males. *J. Econ. Entomol.* 48(4):459-462.

Kono, Tatsuro. 1952. Time-dispersion curve. Experimental studies on the dispersion of insects. *Res. Pop. Ecol.* 1:109-118.

Körting, A. 1931. Beobachtung uber die fritfliege and einiger getreidethysanopteren. *Zeit. angew. Entomol.* 18:154-160.

Krefting, Laurits W. and Eugene I. Roe. 1949. The role of some birds and animals in seed germination. *Ecol. Monographs* 19:270-286.

Laird, M. 1951. Insects collected from aircraft arriving in New Zealand from abroad. Vict. Univ. N.Z., No. 11, 30 p.

Lawson, Francis R., Joseph C. Chamberlin and George T. York. 1951. Dissemination of the beet leafhopper in California. *USDA Tech. Bull.* 1030:1-59.

Legner, E. F. and G. S. Olton. 1969. Migrations of *Hippelates collusor* larvae from moisture and trophic stimuli and their encounter by *Trybliographa* parasites. *J. Econ. Entomol.*

Leigh, Thomas F. and Ray F. Smith. 1959. Flight activity or *Colias philodice eurytheme* Boisduval in response to its physical environment. *Hilgardia* 28:569-624.

LeRoux, e. J. R. O. Paradis and M. Hudon. 1963. Major mortality factors in the population dynamics of the eye-spotted bud moth, the pistol casebearer, the fruit tree-leaf leaf roller and the European corn borer in Quebec. *Mem. Entomol. Soc. Canada* 32:67-82.

Levin, M. D. 1961. Distribution of foragers from honey bee colonies placed in the middle of a large field of alfalfa. *J. Econ. Entomol.* 54:431-434.

Lewis, E. Aneurin. 1949. Tsetse flies carried by railway trains in Kenya Colony. *Bull. Entomol. Res.* 40(4):511-531.

Lewis, T. 1967. The horizontal and vertical distribution of flying insects near artificial windbreaks. *Appl. Biol.* 60:23-31.

————. and J. W. Stephenson. 1966. The permeability of artificial windbreaks and the distribution of flying insects in the leeward sheltered zone. *Ann. Appl. Biol.* 58:355-363.

Liming, O. N., Edgar G. Rex and Kenneth Layton. 1951. Effects of a source of heavy infection on the development of Dutch elm disease in a community. *Phytopathology* 41:146-151.

Lindroth. Carl H. 1953. Some attempts toward experimental zoogeography. *Ecology* 34(4):657-666.

Linford, M. B. 1943. Influence of plant populations upon the incidence of pineapple yellow spot. *Phytopathology* 33:408-410.

Linn, M. B. 1940. The yellows disease of lettuce and endive. *Cornell Univ. Agric. Expt. Sta. Bull.* 742:1-33.

Long, Robert R. 1960. The atmosphere in motion. *Sci.* 131:1287-1292.

MacCreary, Donald. 1941. Comparative density of mosquitoes at ground level and at an elevation of approximately 100 feet. *J. Econ. Entomol.* 34:174-179.

————. and Paul L. Rice. 1949. Parasites of the European corn borer. *Ann. Entomol. Soc. Amer.* 42(2):141-153.

McCubbin, W. A. 1944. Relation of spore dimensions to their rate of fall. *Phytopathology* 34:230-234.

McFadden, E. S. 1941. Stem rust is now migrating to the sunny south to spend the winter months. *Plant Disease Reptr.* 25:24-25.

MacKenzie, Vernon G. 1958. Progress report on air pollution. *U.S. PHS Hlth. Rept.* 73(1):39-41.

MacLachlan, J. D. 1935. The dispersal of viable basidiospores of the *Gymnosporangium* rusts. *J. Arnold Arboretum* 16:411-422.

MacLeod, John and Joseph Donnelly, 1957. Individual and group working methods for fly-population studies. *Bull. Entomol. Res.* 48:585-592.

———— and ————. 1962. Microgeographic aggregations in blowfly populations. *J. Animal Ecology* 31(3):525-543.

MacQuiddy, E. L. 1935. Air studies at higher altitudes. *J. Allergy* 6:123-126.

Markovich, N. Ya. Observations on the dispersal of *Anopheles maculipennis* in connection with the physiological state of the females and the movements of man and domestic animals. *Med. Parasitol.* 10:410-413. (*Rev. Appl. Entomol.* Ser. B. 31:189. 1943.)

Medler, John T. 1960. Long-range displacement of Homoptera in the central United States. XI *Internatl. Congr. f. Entomol.* III:30-35.

Meier, F. C. and F. Artschwager. 1938. Airplane collections of sugar-beet pollen. *Science* 88(2291):507-508.

Minott, Charles W. 1922. The gipsy moth on cranberry bogs. *USDA Bull.* 1093:1-19.

Mischustin, E. von. 1926. Zur Untersuchung der Mikroflora der hoheren Lufschickten. *Centralbl. f. Bakt.* II. Abt. 67:347-351.

Mitchell, Donald F. and Carl Epling. 1951. The diurnal periodicity of *Drosophila pseudoobscura* in southern California. *Ecology* 32(4):696-708.

Moggridge, J. Y. 1949. *Glossinia palipides* Newst. and open country in the coastal area of Kenya. *Bull Entomol.* Res. 40:43-47.

Morlan, Harvey B. and Richard O. Hayes. 1958. Urban dispersal and activity of *Aedes aegypti. Mosquito News* 18(2):137-144.

Morris, Arthur Peebles. 1964. Studies of dispersion of insecticide resistant populations of the house fly, *Musca domestica* L. Rutgers University. Dissertation for Ph.D. p. 1-109.

Morris, K. R. S. 1952. The ecology of epidemic sleeping sickness. II. The effects of an epidemic. *Bull. Entomol. Res.* 43:375-396.

Morris, R. F. and C. A. Miller. 1954. The development of life tables for the spruce budworm. *Canadian J. Zool.* 32:283-301.

————. 1957. The interpretation of mortality data in studies on population dynamics. *Canadian Entomol.* 89(2):49-69.

————. (Editor). 1963. The dynamics of epidemic spruce budworm populations. *Mem. Entomol. Soc. Canada* No. 31:1-332.

Moulton, F. R. (Editor) 1942. *Aerobiology.* Amer. Ass'n. Adv. Sci. Publ. No. 17:1-289. Washington, D.C. 289 p.

Murphy, P. A. 1921. Investigations of potato diseases. *Canada Dept. Agric. Dom. Farms Bull.* 2nd series. No. 44:1-86.

————— and E. J. Worthley. 1920. Relation of climate to the development and control of leaf roll of potato. *Phytopathology* 10:407-414.

Nakamura, Kazuo, Yosiaki Ito, Kazuyoshi Miyashita and Akira Takai. 1964. Dispersal of adult grasshoppers, *Mecostethus magister,* under the field conditions. *Researches on population ecology* 6(2):67-78.

Nash, T. A. M. and J. O. Steiner. 1957. The effect of obstructive clearing on *Glossina palpalis* (R.–D.). *Bull. Entomol Res.* 48:323-339.

Neiswander, C. R. and J. R. Savage. 1931. Migration and dissemination of European corn borer larvae. *J. Econ. Entomol.* 23:389:393.

Neitzel, K. and H. J. Muller. 1959. Erhoehter virusbefall in den Randreihen von Kartoffelbestaenden als folge des Flugverhaltens der Vektoren. *Entomol. Exptl. & Appl.* 2:27-37.

Newhall, A. G. 1938. The spread of onion mildew by wind-borne conidia of *Peronospora destructor. Phytopathology* 28:257-269.

Neyman, J. 1939. On a new class of "contagious" distributions, applicable in entomology and bacteriology. *Ann. Math. Stat.* 10:35-57.

Neyman, Jerzy, and Elizabeth L. Scott. 1957. On a mathematical theory of populations conceived as conglomerations of clusters. *Cold Spring Harbor Symp.* 22:109-120.

————— and ————— 1959. Stochastic models of population dynamics. *Science* 130(3371):303-308.

Nicholson, A. J. 1955. An outline of the dynamics of animal populations. *Australian J. Zool.* 2(1):9-65.

Nielsen, Erik Tetens. 1961. On the habits of the migratory butterfly *Ascia monuste* L. *Biol. Medd. Dan. Vid. Selsk.* 23(11):1-81.

Nielson, M. W. 1968. Biology of the geminate leafhopper, *Colladonus geminatus,* in Oregon. Ann. Entomol. *Soc. America* 61:598-610.

Odum, E. P. 1959. *Fundamentals of ecology.* Saunders Company. Philadelphia. 546 p.

Oertel, E. 1956. Observations on the flight of drone honey bees. *Ann. Entomol. Soc. Amer.* 49:497-500.

Ono, Kosaburo. 1965. Principles, methods and organization of blast forecasting. *The Rice Blast Disease Proceed. Symposium,* IRRI, July, 1963. Johns Hopkins Press.

Öort, A. J. P. 1936. De oogvlekkenzieckteva vi de granen veroorzaakt door *Cercosporella herpotrichoides. Fron. Tidschr. Planzickt.* 42:179-234.

Oosting, Henry J. 1948. *The study of plant communities. An introduction to plant ecology.* W. H. Freeman and Company. San Francisco. 389 p.

Pankiw, P. J. L. Bolton, H. A. McMahon and J. R. Foster. 1956. Alfalfa pollination by honeybees in the Gegina plains of Saskatchewan. *Canadian J. Agric. Sci.* 36(2):114-119.

──────── and C. R. Elliott. 1959. Alsike clover pollination by honey bees in the Peace River region. *Canadian J. Plant Sci.* 39:505-511.

Paradis, R. O. and E. J. LeRoux. 1965. Recherches sur la biologie et la dynamique des populations naturelles d' *Archips argyrospilus* (Wlk.) (Lepidopteres:Tortricidae) dan le Sudouest do Quebec. *Mem. Soc. Entomologique du Canada* 43:1-77.

Parker, J. R., R. C. Newton and R. L. Shotwell. 1955. Observations on mass flights and other activities of the migratory grasshopper. *USDA Tech Bull.* 1109:1-46.

Parker, R. R. 1916. Dispersion of *Musca domestica* Linnaeus under city conditions in Montana. *J. Econ. Entomol.* 9:325-354.

Parlato, S. J., P. J. La Duca and O. C. Durham. 1934. Studies of hyper-sensitiveness to the emanations of the caddis fly (Trichoptera). *Ann. Internal. Med.* 7:1420-1430.

Petersen, L. J. 1959. Relations between inoculum density and infection of wheat by uredospores of *Puccinia graminis* var. *tritici. Phytopathology* 49:607-614.

Peturson, B. 1931. Epidemiology of cereal rusts. *Rept. Dom. Bot.* 1930: 3-5.

Pijl, L. van der. 1969. *Principles of dispersal in higher plants.* Springer-Verlag. Berlin, Heidelberg, New York. 153 pp.

Pimental, David. 1961. The influence of plant spatial patterns on insect populations. *Ann. Entomol Soc. Amer.* 54:61-69.

Plank, J. E. van der. 1947. The relation between the size of plant and the spread of systemic diseases. I. A discussion of ideal cases and a new approach to problems of control. *Ann. Appl. Biol.* 34(3):376-387.

──────── . 1948. The relation the size of plant and the spread of systemic diseases. II. The aphid-borne potato virus diseases. *Ann. Appl. Biol.* 35(1):45-52.

──────── . 1949. The relation between the size of fields and the spread of plant diseases into them. II. Diseases caused by fungi with air-borne spores; with a note on horizons of infection. *Empire J. Exptl. Agric.* 17:18-22.

──────── . 1949a. The relation between the size of fields and the spread of plant disease into them. Part III. Examples and discussion. *Empire J. Exptl. Agric.* 17:141-147.

Pollard, H. N., W. F. Turner and G. H. Kaloostian. 1959. Invasion of the southeast by a western leafhopper, *Homalodisca insolita*. *J. Econ. Entomol.* 52(2):359-360.

Poos, F. W. 1932. Biology of the potato leafhopper, *Empoasca fabae* Harris, and some closely related species of *Empoasca*. *J. Econ. Entomol.* 25:639-646.

Pope, O. A., D. N. Simpson and E. N. Duncan. 1944. Effect of corn barriers on natural crossing in cotton. *J. Agric. Res.* 68:347-361.

Porter, D. R. 1935. Insect transmission, host range and field spread of potato calico. *Hilgardia* 9:383-394.

Posey, G. B. and E. R. Ford. 1924. Survey of blister rust infection on pines at Kittery Point, Maine, and the effect of *Ribes* eradication in controlling the disease. *J. Agric. Res.* 28:1253-1258.

Pound, Glenn S. 1947. Beet mosaic in the Pacific Northwest. *J. Agric. Res.* 75:31-41.

Proctor, B. E. 1934. The microbiology of the upper air. I. *Proc. Amer. Acad. Arts & Sci.* 69:315-340.

_____. 1935. The microbiology of the upper air. II. *J. Bact.* 30:363-375.

Profft, J. 1939. Userfluggewohnheiten der blattläuse im zusammenhang mit der verbreitung von kartoffelvirosen. *Arb. Physiol. Angew. Entomol.* Berlin 6:119-145.

Prokopy, Ronald J., Edward J. Armbrust, Warren R. Cothron, and George G. Gyrisco. 1967. Migration of the alfalfa weevil, *Hypera postica* (Coleoptera: Curculionidae), to and from estivation sites. *J. Econ. Entomol.* 60(1):26-31.

Provost, Maurice W. 1952. The dispersal of *Aedes taeniorhynchus*. I. Preliminary studies. *Mosquito News* 12(3):174-190.

_____. 1960. Current research in mosquito biology and control at Florida's Entomological Research Center. N. J. Mosquito Extermination Ass'n. Proc. 46th Ann. Meet.:64-69.

Quaintance, A. L. and E. L. Jenne. 1912. The plum curculio. *USDA Bureau Entomol. Bull.* 103:1-250.

Rempe, Helmut. 1937. Untersuchungen über der Verbreitung des Blutenstaubes durch die Luftströmungen. *Planta* 27:93-147.

Ribbands, C. R. 1951. The flight range of the honeybee. *J. Animal Ecol.* 20:220-226.

Richards, Merfyn. 1956. A census of mould spores in the air over Britain in 1952. *Trans. Brit. Mycol. Soc.* 39:(4):431-441.

Rittenberg, S. C. 1939. Investigations on the microbiology of marine air. *J. Marine Res.* 2:208-217.

Roebuck, A., L. Broadbent and R. F. W. Redman. 1947. The behaviour of adult click beetles of the genus *Agriotes* (*A. obscurus* L., *A. lineatus* L., and *A. sputator* L.). *Ann. Appl. Biol.* 34:186-196.

Romney, V. E. 1939. Breeding areas and economic distribution of the beet leafhopper in New Mexico, southern Colorado, and western Texas. *USDA Circ.* 518:1-14.

de Ropp, R. S. 1948. The movement of crown-gall bacteria in isolated stem fragments of sunflower. *Phytopathology* 38:993-998.

Rosebury, Theodor. 1947. *Experimental air-borne infection.* Williams and Wilkins Company Baltimore. 222 p.

Rowley, Wayne A. and Charles L. Graham. 1968. The effect of temperature and relative humidity on the flight performance of female *Aedes aegypti. J. Insect Physiol.* 14(9):1251-1257.

Rudolfs, W. 1923. Observations on the relations between atmospheric conditions and the behavior of mosquitoes. *New Jersey Agric. Expt. Sta. Bull.* 388:1-32.

Schmidt, W. 1925. Massenaustausch u. freierluft. *Prod. d. Komischen Phys.* 7:72-77.

Schmitt, C. C., C. H. Kingsolver and J. F. Underwood. 1959. Epidemiology of stem rust of wheat: Wheat stem rust development from inoculation foci of different concentration and spatial arrangement. *Plant Dis. Reptr.* 43(6):601-606.

Schneider, F. 1962. Dispersal and migration. *Annu. Rev. Entomol.* 7:223-242.

Schoof, H. F. and R. E. Siverly. 1954. Urban fly dispersion studies with special reference to movement patterns of *Musca domestica. Amer. J. Trop. Med. & Hyg.* 3:539-547.

———— and ————. 1954a. Multiple release studies on the dispersion of *Musca domestica* at Phoenix, Arizona. *J. Econ. Entomol.* 47:830-838.

Schread, J. C. 1932. Behavior of *Trichogramma* in field liberation. *J. Econ. Entomol.* 25:370-374.

Scudder, Samuel H. 1887. The introduction and spread of *Pieris rapae* in North America, 1860-1885. *Mem. Boston Soc. Nat. Hist.* IV(3): 53-69.

Simons, John N. 1958. Controlling aphid-borne virus diseases. *Sunshine State Agric. Res. Reptr.* July (3):6-7.

Smith, J. E. and G. E. Newell. 1955. The dynamics of the zonation of the common periwinkle (*Littorina littorea* (L.)) on a stony beach. *J. Animal Ecol.* 24:35-56.

Snapp, Oliver I. 1930. Life history and habits of the plum curculio in the Georgia peach belt. *USDA Tech. Bull.* 188:1-91.

Stakman, E. C., A. W. Henry, G. C. Curran and W. N. Christopher. 1923. Spores in the upper air. *J. Agric. Res.* 24:599-606.

_____ and L. M. Hamilton. 1939. Stem rust in 1938. *Plant Dis. Reptr. Suppl.* 117:69-83.

_____ and Clyde M. Christensen. 1946. Aerobiology in relation to plant disease. *Bot. Rev.* 12:205-253.

_____ and J. George Harrar. 1957. *Principles of plant pathology.* The Ronald Press Company. New York 581 p.

Stearns, L. A. and W. R. Haden. 1932. An effective supplementary measure for control of the plum curculio on peach. *Trans. Peninsula Hort. Soc.* 22(5):107-112.

_____, L. L. Williams and W. R. Haden. 1935. Control of the plum curculio in Delaware. *Del. Agric. Expt. Sta. Bull.* 193:1-28.

Steiner, H. M. and H. N. Worthley. 1941. The plum curculio problem on peach in Pennsylvania. *J. Econ. Entomol.* 34:249-255.

Steiner, Loren F. 1952. Methyl eugenol as an attractant for oriental fruit fly. *J. Econ. Entomol.* 45(2):241-248.

Stepanov, K. M. 1935. Rasprostranenle infektsionnykh boleznei rasteni vozdushnymi techeniiami. (Dissemination of effective diseases of plants by air currents.) *Trudy Zaschch.* Rast II. Ser. Fitopat. No. 8: 1-66. (Transl. C. Ziemet.)

Stover, R. H. 1962. Intercontinental spread of banana leaf spot (*Mycosphaerella musicola* Leach). *Trop. Agric.* 39:327-338.

Strehler, Bernard L. 1960. The biology of aging. *Amer. Inst. Biol Sci. Publ.* 6:3-13.

Surtees, Gordon, 1964. Laboratory studies on dispersion behavior of adult beetles in grain. VIII. Spontaneous activity in three species and a new approach to analysis of kinesis mechanisms. *Animal Behaviour* 12:374-377.

Sutton, O. G. 1947. The problem of diffusion in the lower atmosphere. *Quart. J. Roy. Meterol. Soc.* 73:257-281.

Swynnerton, C. F. M. 1936. The tsetse flies of East Africa—A first study of their ecology, with a view to their control. *Trans. Roy. Entomol. Soc. London* 84:1-579.

Tashiro, H. 1966. Intratree dispersal of the citrus red mite, *Panonychus citri* (Acarina:Tetranychidae). *Ann. Entomol. Soc. Amer.* 59(6): 1207-1210.

Taylor, C. E. and G. C. Johnson. 1954. Wind direction and infestation of bean fields by *Aphis fabae* Scop. *Ann. Appl. Biol.* 41:107-116.

Taylor, L. R. 1958. Aphid dispersal and diurnal periodicity. *Proc. Linnean Soc. London.* 169 (session 1956-57): 67-73.

_____. 1960. The distribution of insects at low levels in the air. *J. Animal Ecol.* 29:45-63.

Thomas, I. and E. J. Vevai. 1940. Aphis migration. An analysis of the results of five seasons trapping in North Wales. *Ann. Appl. Biol.* 27:393-405.

Thompson. W. R. 1939. Biological control and the theories of the interactions of populations. *Parasitology* 31(3):299-388.

Thomson, A. L. 1929. Migration of animals. *Encyclopaedia Britannica.* 14th ed. 15:473.

Timofeeff-Ressovsky, N. W. and E. A. Timofeeff-Ressovsky. 1940. Populationgenetische versuche und *Drosophila* I. Zeitliche und Raumliche verteilung der Individuen einiger *Drosophila*-Arten über das Gelände. Z. *i. A. V.* 79:28-34.

———— and ————. 1940a. II. Aktions bereiche von *Drosophila fenebris* und *Drosophila melanogaster.* Z. *i. A. V.* 79:35-43.

———— and ————. 1940b. Quantitative untersuchungen an einiger *Drosophila* populationen. Z. *i. A. V.* 79:44-49.

Tower, W. L. 1906. An investigation of the evolution of the chrysomelid beetles of the genus *Leptinotarsa. Publ. Carnegie Inst. Wash.* 48:1-320.

Turner, Neely. 1960. The effect of inbreeding and crossbreeding on numbers of insects. *Ann. Entomol. Soc. Amer.* 53:686-688.

Tutt, J. W. 1902. *The migration and dispersal of insects.* E. Stock. London. 132 p.

Ukkelberg. H. G. 1933. The rate of fall of spores in relation to the epidemiology of black stem rust. *Bull. Torrey Bot. Cl.* 60:211-228.

Urquhart, F. A. 1966. A study of the migrations of the Gulf Coast population of the monarch butterfly (*Danaus plexippus* L.) in North America. *Ann. Zool. Fenn.* 3(2):82-87.

Uvarov, B. P. 1928. *Locusts and grasshoppers. A handbook for their study and control.* Imperial Bureau Entomol. London. 352 p.

Vinje, James M., and Mary M. Vinje. 1955. Preliminary aerial survey of microbiota in the vicinity of Davenport, Iowa. *Amer. Midl. Naturalist* 54(2):418-432.

Wadley, F. M. and D. O. Wolfenbarger. 1944. Regression of insect density on distance from centers of dispersion as shown by a study of the smaller European elm bark beetle. *J. Agric. Res.* 69:299-308.

————. 1957. Some mathematical aspects of insect dispersion. *Ann. Entomol. Soc. Amer.* 50:230-231.

Waggoner, Paul E. 1952. Distribution of potato late blight around inoculum sources. *Phytopathology* 42:323-328.

———— and James B. Kring. 1956. Use of shade, tent and insecticide in studies of virus spread. *Phytopathology* 46:562-563.

Wakeland, Claude. 1934. The influence of forested areas on pea aphid populations of *Bruchus pisorum* L. (Coleoptera: Bruchidae). *J. Econ. Entomol.* 28:981-986.

Wallace, H. R. 1955. The influence of soil moisture on the emergence of larvae from cysts of the beet eelworm, *Heterodera schactii* Schmidt. *Ann. Appl. Biol.* 43(3):477-484.

———. 1958. Movement of eelworms. II. A comparative study of the movement in soils of *Heterodera schactii* Schmidt and *Ditylenchus dipsaci* (Kuhn) Filipjev. *Ann. Appl. Biol.* 46(1):86-94.

Waloff, N. and K. Bakker. 1963. The flight activity of Miridae (Heteroptera) living on broom, *Sarothamnus ocoparius* (L.) wimm. *J. Animal Ecology* 32:461-480.

Watanabe, Syuzi, Syunro Utida and Toshiharu Yosida. 1952. Dispersion of insect and change of distribution type in its processes. Researches on Population Ecology. *I. Entomol. Laby.* Kyoto Univ. 53:94-108.

Waters, William E. 1959. A quantitative measure of aggregation in insects. *J. Econ. Entomol.* 52(6):1180-1184.

Wehrle, Valerie M. and L. Ogilvie. 1955. Effect of ley grasses (*Ophiobolus graminis*) on the carry-over of take-all. *Plant Path.* 4:111-113. 108.

Wellington, W. G. 1945. Conditions governing the distribution of insects in the free atmosphere. *Canadian Entomol.* 77:7-15.

———. 1945a. Conditions governing the distribution of insects in the free atmosphere. *Canadian Entomol.* 77:21-28.

———. 1945b. Conditions governing the distribution of insects in the free atmosphere. *Canadian Entomol.* 77:44-49.

———. 1945c. Conditions governing the distribution of insects in the free atmosphere. *Canadian Entomol.* 77:69-74.

———. 1954. Atmospheric circulation processes and insect ecology. *Canadian Entomol.* 86:312-333.

Wellman, F. L. 1935. Dissemination of southern celery-mosaic virus on vegetable crops in Florida. *Phytopathology* 25:289-308.

Wells, William Firth. 1955. Airborne contagion and air hygiene: An ecological study of droplet infections. Harvard Univ. Press. Cambridge. 423. p.

Whitten, R. R. 1938. Flight of elm bark beetle. *USDA Bureau Entomol. & Plt. Quar. News Letter* 5(10):13.

Williams, C. B. 1923. Records and problems of insect migration. *Trans. Royal Entomol. Soc. London,* 1923:207-233.

———, G. F. Cockbill, M. E. Gibbs and J. A. Downes. 1942. Studies in the migration of Lepidoptera. *Trans. Roy. Entomol. Soc. London* 92:101-283.

————. 1958. *Insect migration*. Collins. London. 235 p.

Willis, H. R. 1939. Painting for determination of grasshopper flights. *J. Econ. Entomol.* 32:401-403.

Willis, J. C. 1922. *Age and area: A study in geographical distribution and origin of species*. Cambridge Univ. Press. 259. p.

Wilson, E. E. and G. A. Baker. 1946. Some features of the spread of plant diseases by air-borne and insect-borne inoculum. *Phytopathology* 36(6):418-432.

———— and ————. 1946a. Some aspects of the aerial dissemination of spores, with special reference to conidia of *Sclerotinia laxa*. *J. Agric. Res.* 72(9):301-327.

Wit, F. 1952. The pollination of perennial ryegrass (*Lolium perenne* L.) in clonal plantations and polycross fields. *Euphytica* 1(2):95-104.

Wolcott, George N. 1955. Dispersion to the tropics of the spiraea aphid, *Aphis spiraecola* Patch. *J. Agric. Univ. Puerto Rico* 39(1):32-40.

Wolf, Fred T. 1943. The microbiology of the upper air. *Bull. Torrey Bot. Cl.* 70:1-14.

Wolfenbarger, D. O. 1940. Relative prevalence of potato flea-beetle, *Epitrix cucumeris* Harris, injuries in fields adjoining uncultivated areas. *Ann. Entomol. Soc. Amer.* 33:391-394.

————. 1941. Dispersion of *Hylurgopinus rufipes*. USDA Agric. Bureau Entomol. and Plt. Quar. *News Letter* 8(6):8.

———— and T. H. Jones. 1943. Intensity of attacks of *Scolytus multistriatus* at distances from dispersion and convergence points. *J. Econ. Entomol.* 36:399-402.

————. 1946. Dispersion of small organisms, distance dispersion rates of bacteria, spores, seeds, pollen and insects. Incidence rates of disease and injuries. *Amer. Midland Natur.* 35:1-152.

————. 1948. Border effects of serpentine leaf miner abundance in potato fields. *Florida Entmol. Soc.* 31:15-20.

————. 1954. Biology and control of insects affecting subtropical fruits, Avocado. Fla. Agric. Expt. Sta. Ann. Rept. for yr. endg June 30, 1953:290.

————. 1959. Dispersion of small organisms. Incidence of viruses and pollen dispersion of fungus spores and insects. *Lloydia* 22(1):1-106.

————. 1961. Honey bees increase squash yields. *Fla. Agric. Expt. Sta. Sunshine Reptr.* 7(1):15, 19.

————. 1966. Incidence-distance and incidence-time relationships of papaya virus infections. *Plant Disease Reptr.* 50(12):908-909.

————, J. A. Cornell and D. A. Wolfenbarger, 1974. Dispersal distances attained by insects populations of different densities. *Res. Popul. Ecol.* 16. (In press).

Woodbury, Angus M. 1954. *Principles of general ecology.* The Blakiston Company, Inc. New York. 503 p.

Worley, C. L. 1939. Interpretation of comparative growths of fungal colonies growing on different solid substrata. *Plant Physical.* 14 (3):589-593.

Wright, D. W. and D. G. Ashby, 1946. Bionomics of the carrot fly *(Psilia rosae* F.). I. The infestation and sampling of carrot crops. *Ann. Appl. Biol.* 33:69-77.

Wynne-Edwards, V. C. 1962. *Animal dispersion in relation to social behaviour.* Oliver & Boyd. Edinburgh. 653 p.

Yosida, Tosiharu. 1954. The relation between the population density and the pattern of distribution of the rice-plant skipper, *Parnara guttata* Bremer et Gray. Pattern of the spatial distribution of insects. Fifth Rept. *Oyo-Kontyu* 9(4):129-134. (Transl. Robert A. Yamaura.)

Zentmyer, George A., James G. Horsfall and Philip P. Wallace. 1943. Logarithmic-probit relationship of spore dosage and response in Dutch elm disease. *Phytopathology* 33:1121. (Abstract.)

————. 1961. Chemotaxis of zoospores for root exudates. *Science* 133 (3464):1595-1596.

Zinforlin, Mario. 1969. Perception of spatial relationships and pupation delay in fly larvae (*Sarcophaga barbata*). *Anim. Behav.* 17:323-329.

Zoberi, M. H. 1961. Take-off of mould spores in relation to wind speed and humidity. *Ann Botany,* N. S. 25:53-64.

Zulueta, Julian de. 1950. A study of the habits of the adult mosquitoes dwelling in the savannas of Eastern Columbia. *Amer. J. Trop. Med.* 30(2):325-339.

Index

painted lady, 91, 93
Bumblebee
  foraging, 63

Cacao, 144
Caddisfly, 113, 179
Calandra
  granaria, 46
  oryzae, 156
Callodonus geminatus, 85
Callosobruchus chinensis, 180
Campeloma decisum, 141
Capitophorus fragariae, 50
Capsicum frutescens, 142
Carnation, 172
Cedar, western red, 33
Ceratitis capitata, 59, 146, 165
Ceratocystis ulmi, 17, 45
Chestnut blight, 44
Cheumatopsyche, 179
Chironomid, 131
Chloronemic, 177
Choristoneura fumiferana, 40, 199
Circulifer tenellus, 59, 83, 198
Citrus, 165
Clustering, 164
Coccinelid, 83, 166
Cochliomyia macellaria, 51
Cocksfoot, 95
Coleoptera, 137
Colias philodice eurytheme, 118
Conothurus coniperda, 112, 115, 190
Conotrachelus nenuphar, 17, 35
Copegnaths, 81
Coriolis, 63
Corixidae, 141
Corn, 43, 49, 83, 142, 143, 144, 178
Cotton, 48, 144
Crab, Pacific shore, 173
Crayfish, 149, 187
Cronartium ribicola, 45, 47, 91
Cucumber, 148
Culex tarsalis, 134
Culicoides impunctatus, 194
Curculio, plum, 17, 18, 35, 153
Currant, black, 47
Currents, convection, 82

Dacus dorsalis, 159
Danaus plexippus, 46
Darkness, 95, 114
Dendroctonus monticolae, 23

Density, 177
  host, 171
  intraspecific, 187
  level, 23
  quality, 171
Diabrotica vittata, 148
Directional, 64
Disease
  banana leaf spot, 147
  black arm, of cotton, 109
  crowd, 183
    groundnut, 183
    maize, 183
    rosette, 183
    streak, 183
  damping-off, 40
  Dutch elm, 17, 45
  head blight, 109
  leaf roll, 31
  pineapple yellow spot, 171, 184
  potato late blight, 14, 95
  rice, blast-off, 73, 164
  savoy, sugar beet, 32, 174
  sleeping sickness, 112
  take-all of wheat, 40
Disperser
  active, 5, 33, 34
  passive, 5, 33, 122
Dispersion, 2, 24
  agents, 5
  concepts, 1
  density, function of, 179
  discontinuity, 20, 23
  distance, 10
    horizontal, 42, 61, 100
    long, 10, 147
    short, 10
  factors affecting, 15
    inorganic, 27: aidants, 27, 141; barriers, 27; directions, 27, 42, 44; hindrances, 141, 142, 143; localities, 27
    organic, 163: external, 163; internal, 193
  means, 5
  omnidirectional, 42
  patterns, 6
  segmental parts, 21
  terminology, 3
  time, 152
  unidirectional, 42, 46, 55, 59, 61
  vertical, 20, 70, 79, 90